GALILEO'S PLANET

GALILEO'S PLANET

Observing Jupiter Before Photography

by

Thomas Hockey

University of Northern Iowa

Institute of Physics Publishing

Bristol and Philadephia

British Library Cataloguing-in-Publication Data

A catalogue record for this book is available from the British Library.

ISBN 0 7503 0448 0

Library of Congress Cataloging-in-Publication Data are available

Published by Institute of Physics Publishing, wholly owned by The Institute of Physics, London

Institute of Physics Publishing, Dirac House, Temple Back, Bristol BS1 6BE, UK

US Office: Institute of Physics Publishing, The Public Ledger Building, Suite 1035, 150 South Independence Mall West, Philadelphia, PA 19106, USA

Typeset in LaTeX using the IOP Bookmaker macros
Printed in the UK by Bookcraft Ltd, Bath

AAU-1080

Jupiter by Donato Creti, 1711. (Courtesy of the Vatican Museum.)

To the four Walls

Eleanor
Melba
Ruby
Jane

for their love and example

When a man first practically engages in astronomical work, he naturally becomes inquisitive to know the best class of observation suited to the capacities of his telescope. The firmament exhibits a multitude of objects, from the resplendent sun to the faintest nebula; and looking at the clear nocturnal sky whensoever and wheresoever he will, it will be found to display variety enough and beauty enough to satisfy the most fastidious observer ... The planets, it must be admitted, offer the most interesting features for telescopic examination, for the best instruments exhibit the fixed stars merely as luminous points, incapable of amplification by the highest magnifiers; while the planets are at once resolved into large globes of well-defined outline, displaying a variety of markings upon their surfaces ... Jupiter offers in many respects the most attractive features for examination. He forms a distinct system of his own, and is remarkable for his great magnitude, and for the rapidity of his motions. Not only does he present an interesting object for the naked eye ... but as an object for telescopic scrutiny his appearance is magnificent, exceeding in extent of detail that of any other planet ...

William F Denning

Secretary, Observing Astronomical Society, in *Science for All*

1880

Contents

Preface

The lives of astronomers—more than most—are affected by celestial cycles over which they have no influence. Mine is no exception. While I was still in high school, a semialignment of the outer planets allowed the Voyager spaceprobes (from which much of our contemporary understanding of Jupiter comes) to make their 'grand tour'. Thus, when I entered graduate school, the study of the outer Solar System flourished through the efforts of the Voyager scientists. Later, an unknown comet was diverted from its heliocentric orbit by Jupiter. This comet would become Comet Shoemaker-Levy 9 (SL-9) and bring Jupiter again into the spotlight—when it collided with the planet in 1994. Moreover, it is the laws of celestial mechanics that place the Galileo spacecraft, upon which much of our expanding knowledge of the fifth planet rests, into jovian† orbit as I write. I am fortunate to live during a terrifically exciting time in planetary science.

This project is the result of a ten-year relationship with the giant planet. While others peer through mountaintop and space-poised telescopes, my 'observatory' has been dusty library basements and archives. (My greatest occupational hazard is an allergy to book mould!) Yet, there would be little purpose for this volume without the labour of many modern planetary scientists who have shown what a truly 'magnificent planet' Jupiter is[1].

Most of what *I* know about Jupiter comes from Professor Reta Beebe. Had I cited everything she taught me, this text would be filled with awkward 'personal communication' references. On the other hand, any misconceptions presented are my own.

I would like to thank *my* students who have helped with this endeavour over the years. All are (or were) undergraduates at the University of Northern Iowa (UNI): Melinda Carriker, Amy Freiberg, Brian Hynek and Mary Ott. I especially thank my research assistant Sarah Lusson, who aided me in putting the manuscript into final shape.

I heartily acknowledge the translators who read for me the French, German, Italian and Latin I could not: Sylvia Baggett, Andrea Dobson, William Eamon, Anne Hockey, Dawn Ibis, Brad Jacobson, Kevin Krisciunas, Laurie Nicol, Germana Nijim, Erica Nolan, Sheri Pressler, William Sheehan, Chris

---

† The adjectival form for Jupiter, 'jovian', comes from the archaic 'Jove'.

xi

Trayner, Regina Sattler and Kate Seiler.

Thomas Williams and Richard Baum helped me with names and dates. I am indebted to the many authors of the *Dictionary of Scientific Biography*[2] for supplemental biographical information.

Portions of the book have appeared previously in the following.

- Fischer D 1998 *Mission Jupiter: The Spectacular Journey of the Galileo Spacecraft* (Basel: Birkhäuser) at press
- 1996 The Australian pre-discovery of Jupiter's great red spot *Eos, Transactions, American Geophysical Union* **77** (22) W72
- 1996 The search for historical impact sites on Jupiter *Planetary and Space Science* **44** (6) 559
- Fischer D and Holger H 1994 Eine kurze Geschichte der Beobachtung Jupiters *Der Jupiter Crash* (Basel: Birkhäuser)
- 1994 The Shoemaker-Levy 9 spots on Jupiter: their place in history *Earth, Moon, and Planets* **66** (1) 1
- 1993 A brief history of the planetary telescope *Griffith Observer* **52** 10 (2) 16, (3) 11
- 1992 Seeing red: observations of color in Jupiter's equatorial zone on the eve of the modern discovery of the Great Red Spot *Journal for the History of Astronomy* **23** 93
- 1991 Nineteenth century investigations of periodicities in the jovian atmosphere *Vistas in Astronomy* **34** 409
- 1991 Early documentation of the history of Jupiter's South Equatorial Belt disturbance *Bulletin of the American Astronomical Society* **23** (3) 1134
- 1990 An early color rendering of the Planet Jupiter *Bull. Am. Astron. Soc.* **22** (4) 1192
- 1989 A historical interpretation of the study of the visible cloud morphology on the planet Jupiter: 1610–1878 *Publications of the Astronomical Society of the Pacific* **101** (643) 869
- 1989 Planetary photospheres: the story of a 'misdiscovery' *Publ. Astron. Soc. Pacific* **101** (638) 434
- 1989 The first publication of the physical appearance of Jupiter as seen from the United States *Bull. Am. Astron. Soc* **20** (4) 950

Thanks go to those who read and commented on draft versions of this present work: William Sheehan, Audouin Dollfus, Hans-Jörge Mettig, Owen Gingerich, Kevin Krisciunas and other, anonymous reviewers. Their suggestions result in much-improved text.

UNI Deans Gerald Intemann and David Walker provided financial support during the writing process. Nancy Howland, UNI Earth Science Department secretary, produced innumerable photocopies and posted nearly as many letters on my behalf. The assistance of Rosemary Meany, and the Donald O Rod Library staff, in providing materials is much appreciated.

I also would like to thank Ruby Hockey, June Stageberg, Richard Baum, Brenda Corbin, David Levy and others who encouraged me to put this work into its current form. I am grateful to Peter Binfield (editor, Institute of Physics Publishing) for his patience and gentle nurturing of *'Galileo's Planet'*. Finally, a big 'thank you' goes to Wayne Anderson, former Head of, and other colleagues in the Department of Earth Science, UNI, for once upon a time taking a chance on a then twenty-six year old 'Visiting Assistant Professor'.

Thomas Hockey
Associate Professor of Astronomy
UNI
hockey@uni.edu
July 1998

Note: Many pictures of Jupiter, drawn as seen through the eyepiece, are inverted (due to the optics of the telescope) from what has become conventional orientation. To aid comparison, all images of Jupiter that appear in this book are accompanied by an arrow indicating jovian north.

Endnotes

1. The phrase, 'Magnificent Planet', was first used to describe Jupiter in one of the most popular astronomy textbooks of all time: John Herschel's 1849 *Outlines of Astronomy* (Philadelphia: Lea and Blanchard)

2. Gillispie C (ed) 1970–1990 *Dictionary of Scientific Biography* (New York: Scribner)

Glossary

(Definitions of **boldfaced** (first appearance) terms in the text.)

aberration Producing an optical image of an object or scene that is not true to that object or scene.

achromaticity Character of an optical system that counters the effects of dispersion. (Dispersion occurs when different wavelengths of light refract through different angles.) This normally is accomplished by using a compound lens in which the individual elements have different indexes of refraction. They are chosen so that two or more wavelengths come to a focus in the same plane. Alternatively, an 'achromatic' is a refracting telescope in which the coloured fringes resulting from dispersion are eliminated using a system as described above.

air mass An observer looking at his or her zenith observes through '1 air mass'. As zenith angle increases, the observer must look through longer air masses. The atmosphere becomes less transparent, and the effects of bad seeing are amplified.

albedo A quantitative measurement of a surface's reflectivity. 'High' (on a scale of 0 to 1) albedo refers to a highly reflective surface, 'low' albedo to a highly absorbing one.

anticyclonic Clockwise rotation in the northern hemisphere, anti-clockwise in the southern hemisphere.

aperture The diameter of the light-gathering area of a telescope.

apparition For a superior planet, the period between conjunctions with the Sun (the times when the planet is not observable).

astrometry Angular and positional measurement on the celestial sphere.

astronomical unit A unit of distance equal to the mean distance between the Earth and Sun (= 149,600,000 km).

conjunction A planet is either between the Earth and the Sun (inferior planet) or on the opposite side of the Sun from the Earth (inferior or superior planet).

cosmology The study of the structure and evolution of the Universe.

cyclonic Anti-clockwise rotation in the northern hemisphere, clockwise in the southern hemisphere.

cylindrical projection Mapping a sphere onto a cylinder. The cylinder then can be transformed into a planar chart.

declination Coordinate on the celestial sphere. $+90°$ = north celestial pole. $0°$ = celestial equator. $-90°$ = south celestial pole.

diurnal Having a period equal to that between sunrise and sunrise.

ellipsoid A solid figure having the shape of an ellipse in three orthogonal planes.

ephemerides For a celestial object, predicted celestial position at given times. Alternately, an 'ephemeris' is a listing of ephemerides.

focal length The distance behind a lens or mirror at which light from an (infinitely) distant source comes to a focus.

following The side of a planetary feature opposite the direction of the planet's rotation. This is east on Jupiter.

Foucault test Using telescope-maker Jean Foucault's (1819–1868) 1859 'knife edge' technique, opticians can detect deviations from proper curvature smaller than a wavelength of light.

heliocentric 'Sun-centred', especially as in a cosmological model assuming a moving Earth and stationary Sun.

inferior planet Orbits the Sun at a mean distance less than that of the Earth.

insolation The rate of delivery of all direct solar energy per (orthogonal) unit surface area.

limb The edge of a planet's visible hemisphere.

magnification In a telescope, its 'power' (in units of $N\times$). It is calculated by dividing the focal length of the objective lens or mirror by the focal length of the eyepiece.

magnitude A quantitative scale of brightness for celestial objects. $m = 1$ corresponds to a bright star; $m = -6$ corresponds to the dimmest star observable with the naked eye.

maximum angular diameter The apparent size of a planet at minimum distance from the Earth.

micrometer An instrument mounted in the focal plane of a telescope. Usually, a fine line (hair) is moved over a fixed scale, thereby allowing precise angular measurements of objects in the field of view.

objective The light-gathering element in a telescope.

obliquity The angle made by a planet's rotation axis with respect to a line normal to the plane of the planet's orbit. The Earth's $23\frac{1}{2}°$ obliquity is responsible for our world's varying seasons.

occultation The blocking of a source of light by a nonluminous body.

opposition When the Earth is between a superior planet and the Sun. Equivalently, when a superior planet is closest to the Earth.

penumbra The shadow caused by the partial occultation of a light source.

preceding The side of a planetary feature that corresponds to the planet's direction of rotation. This is west on Jupiter.

quadrature The Sun–Earth–Jupiter angle equals 90°.

reflector A reflecting telescope; one that uses a curved mirror to collect light.

refractor A refracting telescope; one that uses a lens to collect light.

second of arc 1/3600th of one degree.

seeing In regard to an astronomical telescope, the steadiness of the Earth's atmosphere through which an observer peers. During bad seeing, variable atmospheric refraction impairs the resolution of the telescope.

sidereal Referring to a measurement made with respect to the Celestial Sphere.

speculum A shiny metal alloy used to make a telescope mirror. Alternatively, a reflecting telescope that uses a speculum mirror.

superior planet Orbits the Sun at a mean distance greater than that of the Earth.

terrestrial Earth-like. (The terrestrial planets are Mercury, Venus, Earth and Mars.)

transit One celestial object moving in front of another object or position. Alternatively, the process of doing so.

zenith The point directly overhead.

Chapter 1

Introduction

Crossing the zone of asteroids on our journey outward from the sun, we meet with a group of bodies widely different from the 'inferior' or terrestrial planets. Their gigantic size, low specific gravity, and rapid rotation, obviously from the first threw the 'superior' planets in a class apart; and modern research has added qualities still more significant of a dissimilar physical constitution. Jupiter, a huge globe 86,000 miles in diameter, stands pre-eminent among them. He is, however, only primus inter pares; all the wider inferences regarding his condition may be extended, with little risk of error, to his fellows; and inferences in his case rest on surer grounds than in the case of the others, from the advantages offered for telescopic scrutiny by his comparative nearness.

Agnes M Clerke, Fellow, Royal Astronomical Society

1885

Why write a book about the planet Jupiter?[1] It has been suggested that an objective catalogue of the contents of our solar system would list the Sun, planet Jupiter, and debris. Yet, as inhabitants of the 'debris', ours is a vista that emphasizes the significance of the Sun and those **terrestrial** worlds (for example, our own Earth) that seem to huddle about it as if for warmth. Clearly, though, the objective description is correct in terms of gravitational influence and sheer volume. Jupiter dwarfs even the other giant planets: Saturn, Uranus, and Neptune. Perhaps, too, the description is correct in terms of the comparative complexity of Jupiter. This includes the planet's rich chemical make-up, unknown physical structure, tremendous magnetic field and myriad of visible atmospheric phenomena.

Much has been written about the Sun. Yet little is said about humankind's interaction with the body that at the time of its formation missed becoming, by a factor of only 100, the second great source of light and heat in our sky, a component of a binary system named Sol.

That interaction was both intimate and rich. Jupiter can be the brightest

1

object in the night sky. It appeared as a large disc through even the crude telescope optics of the 17th century. Moreover, a patient observer could easily exhaust and commit to memory the features that appear on a dead and unchanging world like the Moon. The other planets offered only frustratingly vague detail, or a featureless disc. But Jupiter is bedecked with features! Not just subtle markings these; jovian features approach the garish. Add to this the fact that Jupiter's appearance is constantly changing. An observer can return to its disc repeatedly, always with the prospect of seeing something new. In both hue and intricacy, Jupiter outdoes any other telescopic world and teases, with a wealth of detail, those who would speculate about its nature.

Why write a book about the history of the study of Jupiter? The history of planetary science has been neglected in favour of the stories of celestial measurement, gravitational astronomy and astrophysics. The reason for this is straightforward: emphasis in the history of science mirrors the popularity of the field of inquiry. Planetary astronomy is a distinct observational enterprise apart from the study of the motions of the planets (or, for that matter, astrology). It has, throughout most of its existence, stayed on the periphery of general astronomy. There have been two exceptions to this. One was the initial few decades after the invention of the telescope. This was a time when the other worlds of the solar system—and to some limited extent the Sun—were the only extended objects available for magnified inspection. The other dates from the 1960s and continues today. It coincides with the age of spaceprobe exploration. Planetary astronomy is the only kind of astronomy that can be undertaken *in situ*.

We live in that latter, magic age; it is my intent to remind us of the first (and its inheritance). Specifically, in a day when Pioneer and Voyager spaceprobe images routinely appear in news magazines, I wish to remind people of *two* other periods (that include most of human history). I refer to times when Jupiter and the other planets were mysterious lights in the sky and, later, newly 'discovered' worlds through the eyepiece of that product of the Renaissance, the astronomical telescope.

To do so now is particularly timely. In July 1994, a comet many kilometres across rained down upon Jupiter. Each fragment imparted on that world an explosive energy greater than that of humankind's combined nuclear arsenal. This dramatic event aroused public interest in astronomy like no other in recent memory. Media attention transcended popular-science periodicals and electronic bulletin boards; Jupiter became the focus of articles in major newspapers and discussion on radio and television specials around the world. No less noteworthy a magazine than *Time* pictured Jupiter on its cover[2].

The comet impact of 1994 placed Jupiter in the popular consciousness. Events of 1995–97 caused it to remain there. In December 1995, the Galileo spacecraft completed its long journey to Jupiter. Galileo is the first of a series of 'high tech' spaceprobes capable of imaging the planets as never before. Galileo also released a probe, which descended into the atmosphere of this gaseous world for the first time. Galileo's two-year-plus mission in orbit around Jupiter will provide a steady stream of new information about the giant planet. The late 1990s are the 'years of Jupiter'.

The starting point for this work is the earliest appreciation by humankind of the bright points in the night sky. This date is lost in pre-history.

My ending point is only slightly less arbitrary. The 1880s brought with them both the advent of astrophysics and the professionalization of astronomy. There is a wealth of information available, about nearly every aspect of astronomy, beginning with this decade. In the case of Jupiter, Bertrand Peek's classic *The Planet Jupiter*[3] draws historical notes from this time forward.

Left as my purview, then, is a period where the human eye (not the camera) was the primary astronomical detector. Observers tracked the motions of planets with quadrants or armillary spheres. Later, data were recorded at the eyepiece of a telescope with pen and ink or other drawing tools. Planetary astronomers were largely amateurs, often making scientific progress outside the mainstream of modern science at the time.

The prime resource for this work is the written words of astronomers and theoreticians[4]. I have searched libraries and archives throughout the United States and Canada for these materials. I also have travelled to Europe, the home of most planetary science during the 17th, 18th, and 19th centuries.

Besides text, I have examined more than 200 drawings of Jupiter, made by direct telescopic observation. They span a period from that of Giovanni Cassini to 1898. I have attempted to discover trends in the work of individuals and apparent biases that influenced their way of committing Jupiter to paper. To some extent, I have made an assessment of their abilities as technical artists, as well. I will refer to these drawings in the text where appropriate. Representations that appear to be completely nonphysical (at least from a 1990s viewpoint) I will describe as suspect, with reasons listed. I provide a selection of representative useful drawings as figures.

This is not exclusively an astronomy book. It is a historical examination of a particular science. Therefore, it will not pre-suppose much specialized, scientific training on the part of the reader. Chapter 2 briefly describes Jupiter based on interpretations from 20 years of Pioneer, Voyager, Hubble Space Telescope (HST) and Galileo image analysis. This picture of Jupiter will serve as a reference for the historical discussion to follow. (For instance, specific features and aspects of the planet now recognized as common in high-resolution images of Jupiter will be reintroduced in the chronological context of when they first were observed at low resolution.) Here, too, the system of nomenclature used to describe the geography of Jupiter is explained.

Chapter 3 begins a chronological account of the observations of Jupiter. Historical astronomy records start with the ancient Sumerians, Egyptians and Greeks who recorded the motion of a celestial god across the heavens. All early people's attempts to answer that very human of questions, 'Where is my place in the Universe?', began with interpretations of the moving dots in the sky.

Chapter 4 sketches the accomplishments of 1610 to approximately 1700. I pay particular attention to the debt owed by 17th-century protagonists of the Copernican **heliocentric** theory for supporting evidence provided by observations of Jupiter. No other planet (except, perhaps, the Earth) was so directly involved with settling this major **cosmological** question of paradigm posed during the latter Renaissance.

Chapter 5 documents the transfer of preeminence in planetary astronomy from Italian to German observers. I also present some reasons for the descent of planetary astronomy in general during the 18th century.

I highlight the history of the first morphological phenomena definitely associated with modern features in chapter 6. These observations signalled a transition from dark markings—of the kind used most often for rotation timing—as features of principal interest to bright ones. This transition suggests another: that from primary interest in the measurement of the positions of Jupiter (and features thereon) to interest in the nature of these features themselves. To this parallel list of transitions must be added the continued shift, now from German to English (and American), in predominance in the field.

Circa 1860 was heralded by planetary astronomers beginning to concern themselves not only with improving instrumentation—the use of the reflecting (mirrored) telescope expanded beyond a small group of British enthusiasts during this time, but also the need for *comparative* observations from **apparition** to apparition. I attempt to profile these observers of the mid-19th century (in chapter 7) as a group. I use their backgrounds, observing styles, and instrumentation to suggest systematic biases that appeared in their observation reports. (It is only at this time that the density of reports reached a level that makes such interpretation statistically meaningful.)

Chapter 8 details the modern discovery of the most famous of planetary features, the Great Red Spot. Awareness of this spectacular feature, unlike anything else in the solar system, led to renewed enthusiasm for the observation of Jupiter and indirectly the other planets. Its accessibility, through comparatively small telescopes, created a new generation of amateur planetary observers who greatly influenced observational planetary astronomy well into this century.

Chapter 9 retraces the story of the scientific investigation of Jupiter from the theoretical as opposed to the observational side. Specifically, studies of the physical appearance of the planet naturally led to speculation about its internal identity in terms of composition and structure. Similarly, ideas about the planet's origin and evolution emerged.

Here I tell of the development of jovian models. 'Joviophysics' is an exemplary case study of interdisciplinary dialogue during the parallel growth of observational technique and the development of gravitational physics and thermodynamics.

In chapter 10, I reach some conclusions based on the study presented. A summary formally ends this work. However, the appended list of references is designed to form a thorough bibliography of the literature available from my study period as well as relevant supplemental and interpretive sources.

Endnotes

1. This book is a scholarly heritor of Arthur Alexander's (1896–1971) 1962 *The Planet Saturn* (London: Faber and Faber) and 1965 *The Planet Uranus* (New York: Elsevier)
2. 1994 *Time* **143** May 23
3. Peek B 1958 *The Planet Jupiter* (London: Faber and Faber)
4. An invaluable bibliographical source has been Houzeau J and Lancaster A 1882 *Bibliographie Générale de l'Astronomie* (London: Holland). Jean Houzeau (1820–1888) was Director of the Brussels Observatory.

Chapter 2

A Jupiter Primer

Jupiter Data

Mass $= 1898.8 \times 10^{24}$ kg
Mean distance from the Sun $= 5.20$ **Astronomical Units**
Radius (equatorial) $= 71\,492$ km
Maximum angular diameter $= 46.9$ **seconds of arc**
Bulk rotation period (sidereal) $= 0.413$ d
Obliquity $= 3.1°$
Albedo $= 0.52$

from *The Astronomical Almanac for the Year 1997*

Specific features on Jupiter will be introduced in their historical context. Here I make a few general comments—from the vantage point of the late 20th century—with which to begin our conversation about the giant planet†.

To tell the story of the other inner planets in our Solar System, planetary scientists adopt nomenclature from Earth geography: valleys, volcanoes, etc. For the major outer planets, we borrow from what we know about Earth meteorology. Jupiter's two principal ingredients, hydrogen and helium, remain fluid under a tremendous range of temperature and pressure. A jingoistic astronaut, bent on planting the flag in Jupiter, would descend futilely to the planet's centre—without ever encountering a solid surface.

Yet a thin, opaque, outer layer hides the nature of Jupiter's vast interior from sight. Hydrogen and helium gas are transparent; this layer is thought to be made of clouds formed from trace chemicals. Ammonia ice crystals produce a white cloud at the temperatures and pressures of Jupiter's dry, upper atmosphere. Ascending jovian 'air', containing ammonia, is the likely source of the bright white seen in the 'zones'—see below—and elsewhere on Jupiter.

† A much better modern introduction to the subject is Reta Beebe's 1997 *Jupiter: the Giant Planet* 2nd edn (Washington: Smithsonian Institution Press)

N
↑

Figure 2.1: Voyager spacecraft image of Jupiter. (Courtesy of NASA.)

The chromophores (colouring agents) responsible for the darker, redder portions of the jovian disc are unknown, but also must represent constituents of cloud layers. Infrared measurements imply that the lower albedo clouds are at a warmer (and hence, lower) elevation than the high albedo clouds. Thus, there must be transparent gaps in the uppermost strata that allow us to see lower ones. Here (for example, above the 'belts' mentioned below) we would look for descending jovian 'air', keeping these realms free of freezing ammonia.

Despite what we think we know about a thin exterior shell of Jupiter, most of the planet remains inaccessible to us. Modelling suggests that the doomed astronaut introduced above would encounter domains of increasing temperature and pressure. Jupiter is presumed to be in nearly hydrostatic equilibrium, but it radiates a great deal of heat left over from its initial formation. While the visible Jupiter is very cold, the planet's core temperature may exceed 20 000 °C.

A first-time observer of Jupiter is struck by its banded appearance. The planet is circumscribed by alternating light and dark bands. Traditionally—and arbitrarily, the dark ones are called 'belts' and the light ones 'zones'.

It is assumed that the belts and zones remain confined to certain latitudes by the terrific eastward and westward winds blowing in Jupiter's visible atmosphere (hundreds of kilometres per hour). These winds, in turn, are thought to be a result of Jupiter's rapid rotation, the fastest of any planet in the Solar System.

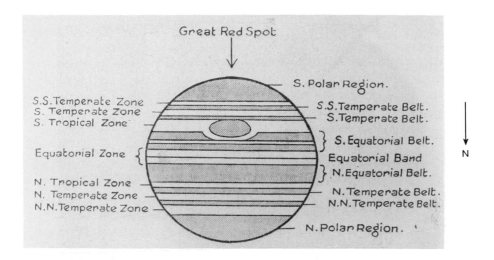

Figure 2.2: Traditional belts and zones of Jupiter. Note that this figure was designed to be used at the telescope; south is up. (Courtesy of John Rogers (British Astronomical Association); unless otherwise noted, all prints were produced by Darrell Fremont (University of Northern Iowa).)

At the turn of this century, a standardized pattern of belts and zones was named. This system was based on that of 19th-century meteorology. The system refers to an idealized Jupiter that does not exist in reality. It assumes static bands at symmetric latitudes. In fact, belts and zones appear and disappear. Light zones become dark, and dark belts turn light. All of the zones and belts can vary in width and albedo.

Still, the system is handy for designating general regions and approximate locations on the planet. The concept of alternating belts and zones was familiar to jovian observers of previous eras; therefore, it is a useful tool in describing early observations.

The definitive source for jovian belt and zone names *circa* 1960 is *The Planet Jupiter*[1] written by British Astronomical Association (BAA) President Bertrand Peek (1891–1961). (Peek was a grammar-school headmaster who operated his own private observatory.[2]) I will adhere strictly to Peek's system with the exception of replacing 'trop'. with 'Tr' as the abbreviation for 'tropical' and deleting periods in abbreviations. In the case of the often variant nomenclature used by astronomers prior to 1900, I have, based on my own interpretation, attempted to assign the more modern terms to the regions referred to.

The BAA/Peek names for the belts and zones (with the abbreviations for them in parentheses) are, from north to south: the North Polar Region (NPR), the North North North Temperate Zone (NNNTZ), the North North North Temperate Belt (NNNTB), the North North Temperate Zone (NNTZ),

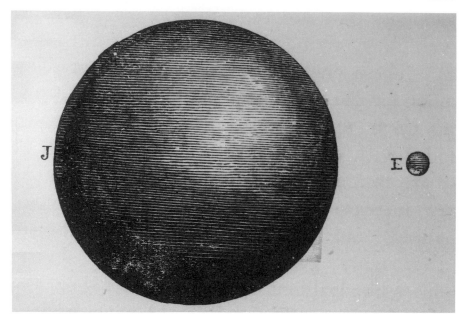

Figure 2.3: The relative size of Jupiter and the Earth. (From R Brown (ed) 1880 *Science for All* (London: Cassell) p 173.)

the North North Temperate Belt (NNTB), the North Temperate Zone (NTZ), the North Temperate Belt (NTB), the North Tropical Zone (NTrZ), the North Equatorial Belt (NEB), the Equatorial Zone (EZ), the South Equatorial Belt (SEB), the South Tropical Zone (STrZ), the South Temperate Belt (STB), the South Temperate Zone (STZ), the South South Temperate Belt (SSTB), the South South Temperate Zone (SSTZ), the South South South Temperate Belt (SSSTB), the South South South Temperate Zone (SSSTZ), and the South Polar Region (SPR). (The polar regions are undifferentiated areas extending to the north and south **limbs**.) Additional belts and zones have been seen from the Earth between the last temperate zones and the polar regions, but these are extremely rare. Indeed, nowhere in the descriptions or drawings of Jupiter examined in this work are the North North North Temperate Zone, South South South Temperate Belt, or South South South Temperate Zone reliably documented.

The major equatorial belts and zones are each divided into northern and southern components, which often behave independently. I have chosen to alter the BAA/Peek abbreviations for these components. For example, instead of writing 'N.comp.EZ' for the northern component of the equatorial zone, I will use parentheses: 'EZ(N)'.

One other feature designated by the BAA is the equatorial band (or EZ Band), a transitory dark streak stationed over the jovian equator. I have avoided more recent bastardized nomenclature for the temporary appearance of bands such as the 'North Equatorial Belt Zone'.

Table 2.1: Approximate latitudinal extents of the major belts and zones (in degrees of planetographic† latitude). (Note that positive latitudes refer to north of the equator and negative latitudes to south of the equator.)

South Temperate Zone	$-42°$ to $-34°$
South Temperate Belt	$-34°$ to $-28°$
South Tropical Zone	$-27°$ to $-20°$
South Equatorial Belt	$-19°$ to $-6°$
Equatorial Zone	$-5°$ to $+6°$
North Equatorial Belt	$+6°$ to $+18°$
North Tropical Zone	$+18°$ to $+23°$
North Temperate Belt	$+24°$ to $+29°$
North Temperate Zone	$+29°$ to $+34°$

I have assigned approximate latitudes to the most prominent belts and zones as they appeared during one recent jovian year[3]. The average historical locations of the bands seem to differ relatively little from these. (See table 2.1.)

Changes in the apparent widths or albedos of the belts and zones can take place quickly. There is reason to suppose that these cloud layers are quite superficial.

Such changes may yield information about physical processes as a function of latitude. However, they do not imply necessarily modification of the jovian wind field (wind velocity as a function of latitude). In fact, an assumption made throughout this work is that the wind field remains quite stable. Evidence for this comes from more than a century of measuring the velocity at which cloud features move across the jovian disc. These measurements were confirmed during the Voyager I and II spaceprobe fly-bys in the 1970s, and by using Hubble Space Telescope images in the 1990s.

At a smaller scale than the belts and zones, meteorological details can be seen in the jovian clouds. The time it takes for a given albedo or colour feature to rotate around Jupiter (its period of rotation) is a function of the planet's rotation rate, plus the velocity of the winds in which the feature blows. As jovian wind velocity varies with latitude, features at different latitudes yield different rotation periods. With only clouds for us to observe, the 'true' rotation period of Jupiter *per se* is masked.

† The more general adjective 'planetographic' is prefered over the Jupiter-specific 'zenographic'. The nonnegligible oblateness of Jupiter has caused two different systems of latitude to be introduced, 'planetographic' and 'planetocentric'. Conversion between systems can be accomplished using

$$\tan(\Theta_{centric}) = (B/A)2\tan(\Theta_{graphic})$$

where B/A is the ratio of the polar radius to the equatorial radius of the planet. $((B/A_{Jupiter})^2 = 0.87458.)$

Jovian atmospheric features include **cyclonic** and **anticyclonic** 'storms', generically named 'spots'. Anticyclonic spots are believed to be regions where material is ascending from a lower level, while cyclonic spots are considered to be regions where material is descending. Liquid brought to a higher, cooler level may freeze; anticyclonic spots often have unique albedos. Aerosols brought to a lower, warmer level may condense; cyclonic spots might represent clear 'windows' on opaque layers below.

Clearly, Jupiter does not reveal its secrets easily, but the 1995 Galileo atmospheric probe demonstrated *in situ* that the basic circumstances of Jupiter heretofore described do, in fact, exist—at least locally[4], and, as most of the planets discovered orbiting other stars in the 1990s bear some resemblance to Jupiter, knowledge of Jupiter leads to a closer understanding of what planets are all about.

Endnotes

1. Peek B 1958 *The Planet Jupiter* (London: Faber and Faber)
2. Cocks E and Cocks J 1995 *Who's Who on the Moon: A Biographical Dictionary of Lunar Nomenclature* (Greensboro, NC: Tudor)
3. Hockey T 1985 *Ground-based Jovian Surface Photometry: 1968–1984.* New Mexico State University Master's Thesis
4. Reports on the Galileo mission appear in the journal *Science.*

Chapter 3

Before the Telescope

Jupiter	*Júpiter*	*Jüpiter*	*Yupiter*	*Jowisz*
Giove	*Guru Brhaspati*	*Kumsung*	*Moku-Sei*	*Mshtarii*
Siupita	*Sui Xing*	*Tzcdck*	*Zeus*	

It was 364 BCE when Jupiter entered the stage of historical record. Chinese astronomer Gan De described it as 'very large and bright'[1]. In the West, it would be centuries later that Ptolemy wrote of the planet[2]. However, Jupiter had long before been acknowledged by humankind.

While not as brilliant as Venus, Jupiter was less fickle. Its position as a **superior** planet meant that its luster did not vary as greatly as did that of **inferior** Venus, which sometimes 'hid' from our eyes in **conjunction** with the Sun.

Venus must always stay close to the Sun in our sky. So it often is diminished by twilight or absent from the night sky altogether. Jupiter is the lord of midnight; it can shine brightly from sunset to sunrise†.

Even remote Saturn changes more in brightness than does Jupiter. All planets shine in reflected sunlight. Because of the long-term variation in the tilt of its rings (with respect to the Earth), the area of Saturn's reflective surface increases and decreases.

The short-term steadiness and constancy of Jupiter's white light—it does not 'twinkle' as much as stars do—earned it the nickname of the 'Night Sun', in the language of the Mayans[3]. The name is apt. On a dark, moonless night, Jupiter may cause objects to cast *shadows*. Some claim to see the planet with the naked eye—in the daytime.

(At its brightest, Jupiter reaches $m = -2.6$ on the astronomers' apparent **magnitude** scale. In comparison, the brightest *star* is $m = -1.4$, and the Full Moon is $m = -12.5$.)

† Superior planet Mars may also outshine Jupiter, but only ocassionally when Mars is particularly close to the Earth. (Houlden A and Stephenson F 1986 *A Supplement to the Tuckerman Tables* (Philadelphia, PA: American Physical Society).)

Figure 3.1: Personification of the Assyrian–Babylonian chief god Marduk, identified with the Roman Jupiter. (From Lehner E and Lehner J 1964 *The Lore and Lure of Outer Space* (New York: Tudor) p 71.)

Jupiter was given the name of the foremost god of the Greek, Babylonian and Scandinavian pantheons (Zeus, Marduk and Thor). Around the Mediterranean, its presumed qualities included moistness, warmth, masculinity and beneficence. The Alexandrian Greeks gave the planet a more descriptive title, Phaethon ('the brilliant one'), but personification eventually won out[4]. The symbol for Jupiter, ♃ , is 'perhaps a rude representation of an eagle, the bird of Jupiter'[5]. Had the ancients had access to even a small telescope, they would have found these names to remain most appropriate. The apparent size of Jupiter's disc is larger than that of the other planets. Also, far from both the Sun and the Earth, it does not appreciably wax and wane as does Venus. If it did, such behaviour might have been deemed unacceptable for a chief deity.

Later, Christianity recast Thursday's planet; it became symbolic for the Judge. On the other hand, Jupiter was also Pride—one of the 'seven deadly sins'[6].

Barrister/author George Chambers (1840–1915) writes

> In by-gone days Jupiter was not without its supposed astrological influences. He was supposed to be the cause of storms and tempests, and to have power over the prosperity of the vegetable kingdom. Pliny thought that lightning†, amongst other things, owed its origin to Jupiter... [7]

In China, Jupiter is associated with wood, spring and the direction East. More importantly, it is the chinese Year Star. Jupiter's apparent march around our celestial sphere marks the passage of the familiar 12-year cycle[8].

Most historical references to Jupiter from before the 17th century have to do, not with the giant planet's intrinsic nature, but with its position and movement in the sky. This scrutiny was in order to solve the cosmological mystery of planetary motion (and was also in the service of astrology). So, for a physical view of Jupiter, we pass quickly through the millennia, to a post-Copernican age (chapter 4).

Endnotes

1. Hetherington B 1996 *A Chronicle of Pre-telescopic Astronomy* (County Durham, UK: Darlington). Records of positional measurements go back hundreds of years earlier.
2. Chambers G 1877 *A Handbook of Descriptive Astronomy* 3rd edn (Oxford: Clarendon)
3. Krupp E 1983 *Echoes of Ancient Skies: the Astronomy of Lost Civilizations* (New York: Harper and Row)
4. Bobrovnikoff N 1990 *Astronomy Before the Telescope* (Tucson, AZ: Pachart)
5. Sharpless I and Philips G 1882 *Astronomy for Schools and General Readers* 3rd edn (Philadelphia: Lippincott)
6. Aveni A 1992 *Conversing with the Planets* (New York: Kodansha)
7. Chambers G *Op. Cit.*
8. Bobrovnikoff N *Op. Cit.*

† The Roman scholar Pliny (23–79) evidently meant to link Jupiter with the source of *meteors*. (Elsewhere, he describes the planet's colour as 'claurus'!) [Bobrovnikoff N 1990 *Astronomy Before the Telescope* (Tucson, AZ: Pachart)].

Chapter 4

The Seventeenth Century

Wonders which great men then discovered at the risk of their lives, now amuse the lighter hours of amateurs... Evidence has been discovered of disturbances going on in those ancient regions [of Jupiter], compared to which terrestrial storms are merest child's play.

Reverend Edward Firmstone, FRAS
1873

As we have seen, humankind has had a long relationship with Jupiter, as one of the great moving lights in the sky. The 17th century, though, was about to catapult Jupiter, and other planets, into a new role. These bodies were about to become worlds in their own right, not just shiny dots to be gazed at, but *places†*.

4.1   First Telescopic Observations

As simply wandering celestial points, the planets lacked unique characteristics, save the colour and brightness of their light, itself borrowed from the Sun. Their motions were a function of their gravity; they may just as well have been uniform billiard balls or abstract mathematical points. And they were often treated as such.

Once their discs were seen, planets could begin to be physically distinguished from the stars. Features on these discs gave the planets personality. Indeed, those who first saw these features were quickly tempted to classify them according to existing schemes. Observers were reminded of the landforms and cloudforms that were familiar to them. Before, the fact that the

---

† Long-time observer John Rogers has written the definitive handbook on the physical observation of Jupiter, *The Giant Planet Jupiter* 1995 (Cambridge: Cambridge University Press). The book describes in detail many of the jovian atmospheric phenomena that I introduce in this work.

Earth was itself a planet was a theoretical construct, based on dynamical arguments. Now that other planets looked a little like the Earth, the similarity between our own and other worlds became more obvious and the Aristotelian conceptual barrier between the Earth and heavens was further broken down.

It was an easy jump from viewing the planetary discs at a distance to imagining the view *from* that other world. Thinking of the planets as places gave them an intellectual approachability. Places, after all, could one day be visited.

The means for this transformation of thought about Jupiter and the other planets was a modest 30-fold improvement in human vision known as the first astronomical telescope. It was fashioned by the great Copernican heretic, Galileo Galilei (1564–1642), a professor at the University of Padua. Using his telescope, Galileo visited the Sun, Moon, planets and Milky Way.

Paradigm shifts in science may take place gradually. Even with historical hindsight it is often impossible to site a specific time and place, when and where a way of thinking about a phenomenon changed. Still, we do know precisely when the telescope was first pointed at Jupiter. In Stillman Drake's translation of the *Siderius Nuncius*, Galileo says dramatically, 'On the seventh day of January in this present year 1610, at the first hour of night, when I was viewing the heavenly bodies with a telescope, Jupiter presented itself to me...'[1]. However, Galileo then hastily proceeds to proclaim his discovery of Jupiter's *satellites* and does not mention the *planet* again!

Galileo can be forgiven. That a planet showed a disc-like, instead of point-like, appearance through a telescope was wondrous, but Jupiter was not unique in this regard. Galileo saw that Venus was also a disc, one which showed phases like the Moon. Saturn's disc was mysterious, exhibiting protuberances that would later be resolved as exquisite orbiting rings. Jupiter's figure was somewhat plain, in comparison.

Galileo did not see features on Jupiter. To do so, his telescope would have had to include a larger **aperture objective**† with a longer **focal length**. Owing to the design of the Galilean telescope, greater power meant a sharply reduced field of view. Galileo's best instrument had a field of view of just over seven minutes of arc[2]. This made it practically suitable *only* for planetary work; even then, without a proper mounting, such a field made pointing awkward at best.

The Galilean telescope did not catch on as a tool for astronomical discovery. With it, only a small number of new astronomical phenomena were documented beyond those already reported by Galileo. Most often, his observations were simply repeated by contemporaries. Jupiter was one of the few objects on the short list of things worth looking at[3].

Francesco Fontana (1602–1656) *did* watch the jovian disc itself. He did so intermittently from Naples between 1630 and 1646[4], when he published a

---

† Galileo's objective lens was only $1\frac{3}{4}$ inches in diameter (Bell L 1922 *The Telescope* (New York: McGraw-Hill)).

Figure 4.1: Galileo's telescopes. (From the frontispiece, Bell L 1922 *The Telescope* (New York: McGraw-Hill).)

book on his astronomical work. (He made the initial report of features on Mars, in 1636[5].) Fontana probably was the first to make regular planetary observations using the improved Keplarian optical design[6].

Fontana was the preeminent telescope-maker in Italy until the 1640s, when his work was narrowly surpassed by that of Galileo's secretary, Evangelista Torricelli (1608–1647). Torricelli is better known for his practical invention of the barometer, the device which led to the conclusion that the atmosphere has weight and that space must be a vacuum.

It would be nice to know who was responsible for first detecting features on Jupiter. Seeing such 'mars' on a planetary disc was a fatal flaw in the perfection of the heavens mandated by Aristotelean cosmology. The Moon was not a uniform radiant disc; it obviously exhibited patches of light and dark. Still, this could be excused in the theory of concentric celestial spheres by placing the Moon on the innermost sphere, closest to the Earth, where it could be contaminated by Earthly imperfection. Jupiter, though, was far from the Earth, supposedly on one of the outer spheres.

Unfortunately, different sources give Fontana, Torricelli or Niccolò Zucchi (1586–1670) credit for first noting the dark belts girdling Jupiter, the planet's most readily apparent features. We do know for certain that Fontana saw as many as three belts in 1633[7], but it is likely that Zucchi saw the NEB and SEB as early as 1630[8].

Zucchi was a Jesuit theologian who eventually achieved high office in Rome. Yet he seems to have had plenty of time for astronomical experimentation. He actually anticipated the reflecting telescope (one that works by the laws of reflection, instead of refraction as Galileo's had) by observing the images of celestial objects reflected in a mirror with a hand-held ocular. There is uncertainty in the historical record as to when he made his observation of Jupiter; we do know that he had carefully studied both Jupiter and Mars by 1640.

In the mid-1640s, Fontana used the belts of Jupiter to hypothesize that the planet must rotate[9]. Another Italian cleric, Francesco Grimaldi (1618–1663), showed the belts to be parallel to the jovian equator in 1648[10]. (Grimaldi, a professor at the University of Bologna, is better known as the discoverer of diffraction[11].)

The first person really to see something *moving* on Jupiter was probably Grimaldi's mentor at Bologna, Professor Giambattista Riccioli (1598–1671). Riccioli studied sunspots, double stars and the planets. A Jesuit, he disapproved of Galileo and his Copernican idea of a moving Earth. (In Riccioli's system of lunar nomenclature, which we have inherited today, Galileo has a very nondescript crater, begrudgingly named for him.)

Riccioli's ambition was to discredit Galileo's interpretation of the Galilean satellites. The new satellites were a danger to geocentricism inasmuch as they seemed a clear case of something revolving around an object other than the Earth. If the four Galileans could orbit the larger Jupiter, why could the five planets (and, for that matter, the Earth) not orbit the supposed larger Sun? However, Riccioli ended up providing evidence in *support* of the heliocentric theory when through his telescope, he made out the actual shadows of the satellite bodies traversing the jovian disc in approximately 1643[12].

In Sicily, father Gioanbatista Odierna (1597–1660) also witnessed jovian satellite shadows in 1652. He used a modest telescope sent to him by Galileo after Odierna wrote a positive review of the *Siderius Nuncius*.

The lack of major discoveries about Jupiter during the first part of the 17th century can be attributed to the fact that the Galilean telescope was still the instrument of choice among astronomers until midway through the 1640s. Indeed, because of its low light-gathering ability, the Galilean telescope was suitable only for looking at bright planets. Even so, the combination of low **magnification** and general image degradation due to **aberration** and imperfect glass-making and grinding techniques made Jupiter's disc an indistinct and often multiple blob much of the time[13]. (Astronomers of the day were most often poor and could not afford the best that the glass-crafting industry could provide, in any case.)

Galileo's own telescope, while crude, probably represented the best of its era. For this reason, and because of Galileo's outspokenness, it was even thought by some that Galileo had found all there was to discover in the sky with the telescope! However, the second half of the century was to bring a second wave of discoveries about Jupiter[14].

4.2 'Recognition'

In discussing features and other phenomena associated with Jupiter, it is important to define explicitly my use of the word 'recognition'. A feature may be observed and actually drawn without it being recognized as a feature. Recognition must include isolation of a specific morphology as something different from its environment. It must be understood to be nonrandom and distinct from other patterns on the planet. Potentially, it should be repeatable. Without recognizing a feature as such, in this way, no feature morphology can be said to have been truly 'discovered'.

Problems arise when a feature is recognized in hindsight to be documented in existing drawings or even in written reports made up of words so carefully chosen to be objective that they do not in themselves call attention to anything. Is such an anachronical discovery valid? I believe not. Even if evidence for a feature exists from a much earlier time, recognition did not take place until something, usually a more modern observation, caused someone to go back and take another look at the older record. The feature may be 'plain as day' on a drawing from the moment it was created, but not 'seen' for decades or centuries afterward. I will call attention to examples of this in my examination of the historical record of Jupiter's appearance.

Discovery, I believe, belongs to the individual or individuals who first recognize a feature *and* (importantly) adequately communicate that recognition to others. (This is a common modern interpretation, and organizational apparatuses exist today designed to give credit based on these two necessary conditions.) I will treat it accordingly. Others have not necessarily done so. I will show examples of challenges to precedence on discovery based on abrogation of the criteria introduced above. Even in our time, disagreements on these points remain. For instance, in deference to other definitions of discovery used regarding that most famous jovian feature of all, I refer to the 'modern' discovery of the feature known as the Great Red Spot (GRS). This did not occur until 1878, when the spot was first recognized, according to my definition, by many who communicated news of its existence to many others.

Figure 4.2: Christiaan Huygens. (From the frontispiece, Bell A 1947 *Christian Huygens and the Development of Science in the Seventeenth Century* (London: Edward Arnold).)

4.3   Sightings of Jovian Features

The legendary Christiaan Huygens (1629–1695) used an extremely long (23 foot) focal length 'aerial' telescope to examine Jupiter. (Its aperture was $2\frac{1}{3}$ inches; it magnified up to $100\times$.[15]) He placed the earliest published drawing of the planet in his 1659 *Systema Saturnium*. (Galileo's Jupiter was merely a circular outline that established the jovian disc but was principally meant to serve as a reference by which to measure the satellite positions.) In it, two equatorial streaks can be seen[16]. This book is more famous, of course, for its accurate description of Saturn's rings.

Some indication of coloured bands parallel to the jovian equator has been evident to observers since Riccioli, at least, but Giovanni Cassini (1625–1712) was the first to describe adequately the belts and zones[17]. Cassini I, as he is sometimes referred to in order to distinguish him from other members of the patrilineal dynasty of astronomers he began, was a student of Grimaldi and Riccioli. His first job was calculating **ephemerides** for an astrologer, but early in his career Cassini took special interest in observing the planets Jupiter and Saturn. He eventually discovered four Saturnian satellites and a major

Figure 4.3: Giovanni Cassini. (Courtesy of Yerkes Observatory.)

division in that planet's rings†. (NASA's recently launched Saturn-orbiting spaceprobe is named Cassini.)

Cassini was also the first to note Jupiter's oblateness (out-of-roundness), which he estimated to be 14/15ths; that is, Jupiter's equatorial diameter was 1/15th greater than its polar diameter[18]. This was confirmed by the Danish astronomer Olaus Römer (1644–1710) and others[19]. (Indeed, Jupiter's oblateness continued to be 'rediscovered' by observers for the next century[20].) These measurements verified the prediction computed by none other than Isaac Newton (1642–1727), who had considered what would happen to a hypothetical liquid water planet as it spun on its axis[21].

Cassini's friend and instrument maker, Giuseppe Campani (1635–1715), was also a planetary observer, during an era when little except a keen eye and the ability to operate a telescope was necessary for this avocation, and the optician was often on equal footing with the astronomer. Such craftsmen had the opportunity to 'test' new instruments and thereby see things that had not been observed before. This is why we see that many discoveries have been attributed to individuals who were primarily telescope-makers. (A more modern example is Alvan Clark's first sighting of the first known white dwarf

† His 'discovery' of a satellite orbiting Venus turned out to be specious. (Baum R and Sheehan W 1997 *In Search of Planet Vulcan: The Ghost in Newton's Clockwork Universe* (New York: Plenum)).

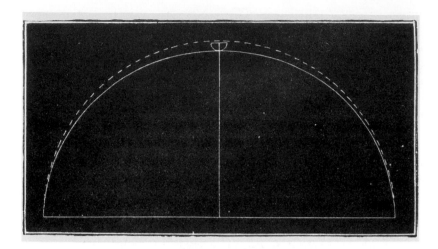

Figure 4.4: The 'polar flattening' of Jupiter. (From Ledger E 1882 *The Sun: Its Planets and their Satellites* (London: Edward Stanford) p 297.)

star, Sirius B, while testing a telescope in the yard of his optical shop.) Conversely, the optician's observing time, even if he was so inclined to make use of it, was limited by always pressing business concerns. Thus, information to be gleaned from long-term observation was most often obtained by the professional astronomer.

According to Albert Van Helden[22], Campani was '...by far the best telescope maker of the second half of the seventeenth century'. Besides Cassini, Campani found a patron in Archduke Ferdinand of Tuscany. He received the attention of royalty, not for his excellent telescopes or the improvements he made in microscopy, but rather for his co-invention of a silent clock for Pope Alexander VII.

Upon looking at Jupiter, Campani discerned structure in the belts and zones: He '...affirms he hath observed by the goodness of his Glasses, certain protuberancies [*sic*] and inequalities, much greater than those that have been seen therein hitherto...'[23]. Campani the scientist went on to study these irregularities to see if they would move, and so betray the fact that Jupiter really did rotate on its axis. Cassini had detected the motion of similar but fainter features by now; however, the time it took for them to make one complete rotation remained uncertain[24].

Planetary rotation was another Copernican trait. In the heliocentric system, the apparent rotation of the entire celestial sphere was caused by the planet Earth's rotation. While less controversial than the Earth's revolution around the Sun, the rotation of the Earth is not obvious from its surface. Observation of this behaviour in another world further opened the possibility of it in our own.

The easiest way to prove planetary rotation was to find on its disc a feature with little longitudinal extent (a 'spot'). This feature could be observed when it was on an imaginary line dividing the planet into east and west halves. Such a line is called the central meridian. If the planet rotated, and the axis of rotation were roughly perpendicular to the plane of the ecliptic, the spot could be observed to move off the central meridian and disappear around the limb of the planet, only to reappear on the opposite limb some time later. Timed observations of two successive central-meridian **transits** (easier to observe than the appearance or disappearance of a feature at the limb) would yield the rotation period of the planet, the length of its 'day'.

Priority for the earliest observation of an intrinsic spot on Jupiter went to either Cassini (along with Campani) or Robert Hooke (1635–1702). The first strictly scientific journal in English, the *Philosophical Transactions of the Royal Society*, appeared in print just in time to document the controversy in volume 1. It staunchly supported Hooke, a British hero of sorts for his work with Christopher Wren in the rebuilding of London, and, in an age when editorial fairness was a new idea, reported Cassini's work only in passing.

Hooke found 'on the ninth of May, 1664, ... a small Spot in the biggest of the 3 obscurer Belts of Jupiter'[25]. Through his 12-foot focal-length† telescope he saw that '...within 2 hours after, the said Spot had moved from East to West, about half the length of the Diameter of Jupiter'. Actually, Cassini only claimed to have seen two 'spots', which were the shadows of Galilean satellites, on July 30, 1664[26]. He suggested that other astronomers look for them, too[27].

Debate about what constituted a true spot was enhanced by an account by Adrien Auzout (1622–1691) suggesting that the observations of Cassini and Campani might not be satellite shadows but rather jovian 'surface' features. To avoid future ambiguity, Auzout proposed looking for a bright body against the primary disc whenever a dark spot was seen. The correlation of these observations would indicate a satellite and its shadow. He thought that the satellite Ganymede might be a particularly noticeable body projected onto Jupiter because of its supposed size and brightness[28]. Unfortunately, detecting a satellite against the bright face of Jupiter was to remain difficult, and this method could not be relied upon to differentiate between an intrinsic spot and a shadow.

Auzout—a Frenchman trying to arbitrate between English and Italian astronomers—later agreed that Cassini had seen satellite shadows. However, now Campani believed the spots to be intrinsic and boasted of his telescope's ability in support of his stance. Auzout eventually conceded that the astronomers from Bologna and Rome had seen *both* satellite shadows and intrinsic spots[29].

† English units of measurements were used almost exclusively by the sources cited in this book. They have been retained in the historical chapters in order to maintain continuity.

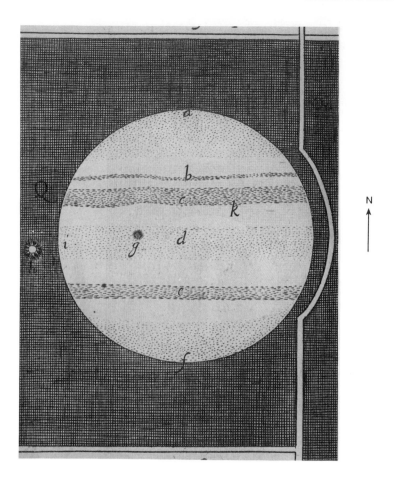

Figure 4.5: Jupiter, 26 June 1666, by Hooke. (From a plate accompanying 1666 *Philosophical Transactions of the Royal Society* **1**; courtesy of the Royal Astronomical Society.)

In any event, Cassini's sighting of a 'permanent spot' in 1665 was the more important observation because the regular recurring passage across the jovian disc of this easily observed feature not only offered proof of Jupiter's rotation but allowed Cassini to calculate the first reliable rotation time for the planet: 9 h 56 min[30].

Hooke missed the opportunity to make this astronomical contribution, just as it appears he fell short of making other scientific discoveries such as Boyle's Law and the inverse-square relation for gravitation. Hooke even overlooked the flattening of the jovian disc, though he was a proponent of the theory that the Earth is an oblate spheroid[31].

4.4 Cosmological Implications

Let us return to the 17th century. As I suggested before, the idea that Jupiter rotated seemed likely to good Copernicans but was not assumed. Furthermore, the suspicion that it did was based only on an analogy: Johannes Kepler (1571–1630) had said that since the planets revolve around the Sun, the Sun should rotate. Kepler's pre-Newtonian theory of planetary motion held that some sort of magnetic-like eminence from the Sun propelled the planets[32]. Thus, the Sun could be likened to a rotating hub pulling the planets transversely. The Sun's rotation was later observationally confirmed, and so Jupiter-observers, who liked to identify the jovian system and its moons as a scale model of the solar system, also believed that Jupiter should rotate because of the satellites orbiting it[33]. Cassini's quantitative proof finally put the matter on firm scientific ground. Ironically, Cassini himself was a conservative when it came to the structure of the solar system. He resisted Kepler's elliptical planetary orbits, preferring the perfect circles of Aristotle. Cassini favoured a model that in many ways resembled the Earth-centred one of Tycho Brahe (1546–1601).

Jupiter was the first planet to have its rotation confirmed. Now that there were two such rotating bodies in the solar system (Jupiter, and, evidently, the Earth itself)—and both had satellites—it was hoped that Saturn with *its* recently discovered satellite (Titan, discovered by Huygens in 1656) also would be shown to rotate. This inductive approach was successful, and soon it was supposed that all planets, regardless of the presence of satellites, were rotating. Said Cassini

> This Observation ought to excite all Curious persons to endeavour the perfecting of optick [*sic*] glasses, to the end that it may be discovered, whether the other Planets, as Mars, Venus and Mercury, about whom no Moon hath as yet been discovered, do yet turn about their axes, and in how much time they do so; especially Mars, in whom some Spot is discover'd [*sic*]...[34].

At the same time, the satellites themselves were not expected to rotate, this time by analogy with the Moon. In the literature of the day there was no discussion of the idea that a revolving body that always presents the same face to its primary *must* rotate.

4.5 Questions on Precedence

A curious note now appeared in the *Philosophical Transactions*:

> Eustachio de Divinis (faith the Informer) has written a large Letter, wherein he pretends, that the Permanent Spot in Jupiter hath been first of all discovered with his Glasses; and that the P. Gotignies is

the first that hath thence deduced the Motion of Jupiter about his Axis; and the Signior Cassini opposed it at first; to whom that said Gotignies wrote a letter of complaint thereupon[35].

Eustachio Divini (1610–1685) was the 'other' optician in Rome and a rival of Campani. His career paralleled Campani's in that he also manufactured clocks, advanced the art of microscope-making and produced long telescopes for, among others, Cassini. As a planetary observer, though, he seems to have been a poor third to Cassini and Campani amidst Italian astronomers.

After Divini's conflicting claim to the discovery of the 'permanent spot', Henry Oldenburg (1618–1677), the editor and translator for the *Philosophical Transactions*, made clear by his tone *his* opinion of such dissent: 'The same Eustachio pretends likewise...'[36]. A prominent footnote states that the spot was first observed in England by Hooke, a founding member of the Royal Society and co-secretary with Oldenburg. 'England' is set in particularly large type.

Yet 'Hooke's spot' was not the 'permanent spot' of Cassini. The descriptions of these two features do not match. The inadequate consideration of the possibility that *two* prominent spots might have appeared within a relatively short period, and been discovered independently by astronomers of two different nationalities, added considerably to the confusion[37].

Cassini's permanent spot was situated next to the STB[38] and '...its diameter is about the tenth part of that of Jupiter; and at the time that its center is nearest to that of Jupiter, it is distant from it about the third part of the semi-diameter of that Planet'[39]†. Hooke's spot was by his own admission small; and, because the NEB was drawn as the most conspicuous belt during the 1660s by all participants in the story[40], it could be placed there. Elongated 'barges' do appear at these latitudes. Extremely dark, they might be nearly as easily resolved as satellite shadows. (See chapter 5.)

4.6   Late Seventeenth-Century Work

By 1666, Cassini had succeeded in using his 17 foot Campani-built telescope to identify the shadows cast by all four Galilean satellites as well as their bodies in transit (the more difficult task). These events were substantiated by Auzout, Huygens and others using a satellite ephemeris provided by Cassini. He also had produced further evidence of jovian rotation such as longitudinal variation in belts and overall changes in disc brightness over short periods[41].

Cassini elucidated how he tried to differentiate between satellite shadows and spots[42]: a satellite shadow moves over its planet's disc at a nearly constant rate determined by the secondary body's orbital period. Cassini noticed that the travel time of a spot *on* Jupiter appears to vary. This is because the

---

† Cassini recognized that the spot appeared smallest when it was viewed obliquely, near the limb. (Cassini J 1740 *Élémens d'Astronomie* Paris.)

velocity of the spot transverse to the observer is greater at the central meridian (where it has no radial velocity) than near the limb (where much of its velocity is radial).

The axis of rotation heretofore had been taken to be parallel to a line drawn perpendicularly to the ecliptic. Cassini now saw that the bands were gently curved and concluded from this that Jupiter's axis might be slightly oblique. Already, it was assumed that the belts and zones were orthogonal to the axis[43].

While Cassini continued to work on Jupiter's rotation, Hooke wrote the first thorough description of the jovian disc. Doing so, he established a pattern that would be followed more or less for over 300 years[44]. This pattern consists of a report on morphology and albedo compiled latitudinally starting from the poles[45]. His instrument was a 60 foot focal-length **refractor**.

Hooke found both poles fairly dark. He identified a northern narrow belt, probably the NTB, as a 'small dark Belt' and a 'great black Belt (NEB)†[46]'. Hooke then wrote of a 'pretty large and bright Zone; but the middle ... was somewhat darker than the edges.' This would have been the EZ. Note that Hooke happened to use the terms 'belts' and 'zones' in a manner not inconsistent with current nomenclature.

Such descriptions are of more help than the earliest drawings. Hooke's accompanying picture of Jupiter[47] is a case in point. The wind jet that marks the northernmost extent of the NTrZ (+23°) is an excellent fiducial mark because of its invariance in latitude.‡ (It is a major inflection point in the jovian wind velocity–latitude plot.) This feature seems present in the Hooke drawing, but if it is accepted to be at the proper latitude, no other recognizable belt or zone can be deduced. Similarly, it is difficult to match Cassini's later drawing with any recognizable pattern of bands[48].

The date of the above observations was 26 June 1666[49]. Hooke did little work at a telescope in later years and concentrated on the science of microscopy. His greatest service to astronomy was not any particular observation, but the invention of the cross-hair telescopic sight. (Auzout later perfected this device.)

Cassini, now at the Paris Observatory, wrote more on the two kinds of 'spot' that he had studied over six years after the appearance of the permanent spot (still credited to Hooke in the *Philosophical Transactions*). By now Cassini was using his satellite ephemeris to determine whether a spotlike darkening was caused by a Galilean satellite[50]. Cassini wanted to see if the potential shadow moved east to west at the rate of one of the satellites (now well known) and whether it preceded where the satellite should be before **opposition** and followed it after. This difference should increase the further Jupiter is from opposition.

† Here and elsewhere, bracketed comments were inserted by the author.
‡ For instance, a ground-based measurement made over 100 years before Voyager I and II placed the jet at +22.55° (Lohse O 1873 Jupiter *Beobachtungen Angestellt auf der Sternwarte des Kammerhern von Bülow zu Bothkamp* **2** 51).

Intrinsic spots do not correlate with the satellites in any way. Cassini likened these to the 'spots' on the Moon (craters). These spots also moved east to west, but '...they never are so well seen as when they approach to the Center, they being very narrow and almost imperceptible, when they approach to the Circumference: which makes us believe, that they are flat, and superficial to Jupiter'[51].

Cassini was the first to produce an ephemeris for jovian spots in 1665[52], but already by the following year it became clear how difficult a task this was to become. Once Jupiter 'got free of the sun-beams [*sic*]', that is, after conjunction 1667, the permanent spot was difficult to make out[53]. Cassini thought that it was disappearing, like sunspots do, and gave up on tracking the feature altogether. However, in January 1672 he saw the spot again 'adhering' to the 'Southern Belt' just as before[54]. Cassini measured its position and attempted to calculate a refined rotation period for the planet. In theory, the span of six years should have allowed him to fix a period with great precision as he assumed that the motion of a spot could not possibly *change* during this interval. The only problem was that over so long a time there existed an uncertainty over the integer number of rotations that the spot had undergone: 5294 or 5295. Thus, Cassini was left with two interchangeable periods, each with a high number of significant digits but differing from one another: 9 h 55 min 58 s or 9 h 55 min 51 s. Because both times were near Cassini's original estimate, the first Cassini ephemeris proved to be not too far off.

Cassini continued to watch the permanent spot and 'fine-tune' his ephemeris, now with a Campani 34 foot. On 1 March 1672 he was for the first time able to observe two consecutive transits of the feature, one early in the evening and one before sunrise the next morning. (Before this, he had always had to lose a transit to daylight.) With these observations in hand, Cassini invited two prominent astronomers to his observatory to witness the spot cross the Central Meridian at the exact moment predicted by Cassini[55]. Cassini now had so much faith in the tenure of the permanent spot and his ability to predict its position accurately that he ventured that the spot could be used as a chronometer to find a navigator's longitude on the Earth!

The determination of longitude was the *raison d'être* for many an employed astronomer. The wealth of shipping companies and the fates of navies rested upon it. The determination of latitude at sea was a straightforward matter. But longitude required an accurate measurement of time, and no shipboard clock of the era was up to the task. If some celestial cycle of the right duration could be documented, it might act as a natural clock in the sky. Many such cycles were tried, including the positions of the Galilean satellites, which arranged themselves in different orders with respect to Jupiter over periods of days and weeks. Better, a transit measurement of a large jovian spot could easily be made by a small telescope, even on the deck of a ship. With an almanac of predicted transit times in hand, a navigator could provide himself with time resolution on the order of minutes, thereby improving the accuracy of his longitude calculation accordingly.

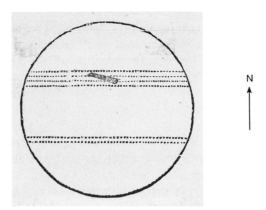

Figure 4.6: Jupiter, 29 May 1686, by G Cassini. (From 1686 *Journal des Sçavans* 295; courtesy of Woody Sullivan (University of Washington) and the Special Collections Division, University of Washington Libraries.)

Yet Cassini's conservatism always led to ever-so-slight inaccuracies in his predicted positions and timings for both spots and the Galilean satellites: Cassini never acknowledged that the velocity of light is a finite constant (as determined in Paris by Römer) and did not introduce the appropriate light-travel time correction for the varying distance to Jupiter. Thus Cassini's transit times were either early or late, depending upon how far Jupiter happened to be from the Earth. Had his plan been implemented, the navigator would have run aground.

The permanent spot vanished in 1674 but reappeared for a time in 1677; from 1685 to 1687; and in 1690, 1692, 1693 and 1694. Each appearance was considered to be a new formation. As the spot was spawned always at the same latitude and– as best as could be determined—the same longitude, Cassini believed that there was a particular tract on Jupiter with a proclivity for growing (not so) permanent spots[56]. More recently, some astronomers have proposed that the permanent spot was an early manifestation of the Great Red Spot. (See chapters 5 and 8.)

Cassini discovered other spots on Jupiter that he considered 'permanent'. In 1686, for instance, a portion of 'la bande la plus large' was very dark[57]. Cassini chose to consider the region as a 'spot' occupying about one-sixth of a jovian diameter. He calculated its rotation time to be within a minute of other spots.

The 'largest band' lent ambiguity to the location of this feature, though an accompanying diagram showed it to be in the NEB. Because of changes on Jupiter and improvements in resolution to come in the near future, such crude means of description were destined soon to fail.

An SEB spot appeared in 1690 with an area 'autant à peu près qu'en occupe toute l'Affrique' [nearly as vast as the whole of Africa][58]. This spot had a rotation period distinctly different from that of the permanent spot:

9 h 51 min. (It, too, changed its shape.) Two EZ spots, seen in 1690–91, had periods of 9 h 52 min 30 s. Cassini was ready to concede that the movement of spots on Jupiter had two components: one of the planet rotating on its axis and one unique to the spots. He worried that an 'infiniti' of observations might be required to separate the two[59]! (Maybe, rationalized Cassini, the fast spots close to the equator were somehow propelled by greater exposure to direct sunlight.)†

There can be little doubt that Cassini was more interested in the orbits of the Galilean satellites than in markings on the planet itself. Presumably, many observations of spots were made incidental to those of Io, Europa, Ganymede and Callisto. Such was the case in 1699, when Cassini happened to see a dark jovian spot while monitoring a satellite becoming eclipsed by its primary. Soon, he noticed a second, similar spot. Jupiter itself became the focus of attention when he spotted a third. This was a new record for the number of large jovian spots visible on a given night[60].

Previously, Cassini had believed that spots were more plentiful when Jupiter was near perihelion (closest to the Sun). However, this veritable outbreak of big spots occurred when the planet was halfway between perihelion and aphelion (furthest from the Sun). The question of spot periodicity was left hanging. '...ce qui nous aprend qu'il y a dans la nature des changemens [sic], dont on nefçauroit s'apercevoir qu'aprés un nombre d'annés', advised Cassini, 'quelquefois plus grand qu'il n'eft accordé à la vie d'un homme' [we therefore learn that there are changes in nature that we could find out only after many years ... sometimes longer than what is granted to a man's life][61].

Cassini was the first to describe dark spots on the order of the size of the Comet Shoemaker–Levy 9 impact complexes. For this reason, another Cassini spot has attracted recent attention. In a December 1692 series of drawings, a dark spot is shown spreading out in the jovian winds. Isshi Tabe, Jun-ichi Watanabe and Michiwo Jimbo suggest that its behaviour is akin to that exhibited by the 1994 spots, caused by Comet Shoemaker–Levy 9's collision with Jupiter[62]. However, there are several other Cassini drawings, with different dates, that could be interpreted similarly[63]! As we will see, intrinsic jovian spots do evolve in the jovian winds. Thus, there is no obvious way to distinguish any of Cassini's spots from the atmospheric spots we see today. The search for exogenic spots must be based on more than just morphology. (I will have more to say on this subject in chapter 10.)

By late in his career, Cassini had seen up to seven or eight 'bandes' on Jupiter[64]. He explained how jovian bands seem to grow wider and thinner, to break apart and rejoin, to appear and disappear. The apparent changing latitude of bands was not caused by Jupiter's obliquity (tilting of the planet's poles toward and away from us each jovian year)[65]. Cassini was convinced that it was real.

---

† Cassini's son, Jacques (1677–1756) thought that spots might move faster when Jupiter was closest to the Sun. (Cassini J 1740 *Élémens d'Astronomie* (Paris).)

Only the equatorial belts seemed fairly constant. Cassini proposed that they were fixed layers of clouds condensed over glassy jovian seas. Dark spots were inland seas, bright spots patches of new fallen snow[66]. It was a pretty, if entirely incorrect, picture.

4.7 Patronage

The birth and early growth in Italy of what would come to be called planetary science followed the first publicized use of a telescope for astronomical discovery there. It might never have achieved the impetus it did if Italians Galileo, Campani and others had not been endowed with the ability to be good craftsmen in the grinding and polishing of optical lenses.

Beyond this, though, was the support system of patronage, familiar to this part of the world, which permitted planetary astronomers to have the necessary equipment, and access to publishing apparatus, that they needed. It also gave them the financial independence required to devote time to the new science. In 17th-century Italy, patronage was all powerful. Even faculty positions and subsequent raises were dependent upon patronage in that they were usually offered only after the intercession of some man (or occasionally woman) of public prominence, on behalf of the academician[67].

Observational astronomy was an excellent profession for generating patronage. It was not considered beneath a scientist of the time to make and sell instruments (e.g., Campani) or supplement his income by giving private lessons. A shrewd astronomer (like Galileo) would forgo the immediate remuneration from marketing some of his instruments and skills and instead offer them as gifts to potential patrons. A telescope made the perfect impression-making gift. It was 'all the rage' in the early 1600s. However, while it did not require expensive raw materials, only certain individuals could construct a workable one. Also, the telescope was often just beyond the ability of a prince or potentate to use without instruction. The astronomer/instrument-maker/gift-giver could now make himself available to accompany his offering and to give a few private lessons on how to operate it[68]. Thus, the astronomer who could make a telescope had a ready and, usually, immediately accepted 'calling card' at his disposal for introducing himself to, and having the ear of, selected luminaries. (The instrument itself, of course, was rarely put to scientific use by the recipient; its purpose was mainly amusement.)

Another powerful technique of ingratiation, available to the astronomer, was to ask permission to dedicate published observations to (and by that strike up a correspondence with) the potential patron. Such requests were seldom turned down. If very lucky, an astronomer—again, Galileo is a fine example—could discover something in the sky and then offer to name it after a dignitary[69]. (The Galilean satellites were first the Medician satellites, after the brothers de' Medici.) While artisans could dedicate *objets d'art* to persons in high places or, more obviously, fashion such a work after the individual (explicitly or symbolically), only the astronomer could immortalize someone in the permanence of the very heavens themselves. No statue, no matter how exquisite, could equal a world as a lasting monument. Even the pyramids will eventually crumble.

By maintaining a flattering exchange after such gestures and keeping abreast of the targeted personage's particular attempts at having scientific interests, no matter how vague, shallow or fickle they were (it was considered good form for the movers and shakers of the 17th century world to have some sort of intellectual bent or, equally, to be perceived as having one), an Italian scientist of merit could be reasonably confident of receiving some tangible support from one major or several lesser patrons. He must only 'keep up' the production of good scientific works in the name of his benefactor, in order to continue to be supported—often the catch[70].

Note that the mechanism outlined above—and no one, except, maybe, Galileo, used all of it so well—could operate in ecclesiastical circles in addition to secular ones. Clerical astronomers brought credit to their church as long as they did not interpret their discoveries in a way contrary to doctrine. The more apparent it became that the planetary observations being made constituted evidence refuting dogma, *vis à vis* my earlier discussion of Copernicanism, the more ordained and devout astronomers avoided explicit interpretation. For a while, an official interpretive eye was closed to the damning observations, but this could not continue for long. A secure church could tolerate fringe unorthodoxy. A church increasingly under attack from Protestantism could not.

In a strictly hierarchical society, there was the opportunity for a relationship to exist between those who lived by their external assets, and those who lived by their mental ones, to exist. The commodity of exchange was prestige. Still, the relationship was more symbiotic than parasitic, and so far I have presented patronage in a neutral light. Its worst aspect yet cited is that a few of the finite number of good astronomical telescopes were kept out of the hands of people who could make good use of them. The real 'down side' of patronage was the need to guard jealously discovery so that proper credit could go to the proper patron[71]. The arguments over precedence already mentioned not only involved personal acclaim, they probably involved perpetuating sustenance.

4.8   A Century of Discovery

The first century of jovian observation was characterized by a busy period of rapid discovery by a small group of telescopic tyros and then a much longer period of low activity while technology caught up with the demand for greater resolution. The beginning few decades of physical planetary astronomy saw the first views of the disc of Jupiter, the belts and later their orthogonal geometry with respect to the rotation axis of the planet, satellite shadows transiting the disc, then larger intrinsic features and later intrinsic spots suitable for rotation-timing. These were the events that made possible the first simple inquiries of any kind into Jupiter's physical nature. Yet after these initial and relatively easy discoveries had been made, Jupiter slowed in giving up spectacular secrets. Subsequent generations were left for a long time largely refining the documentation of the phenomena noticed in those hectic start-up years after the revelation of the telescopic sky.

4.9 Summary

In this chapter I introduced the first telescopic observations made of Jupiter and the initial discoveries about the appearance of the giant planet's disc that quickly followed. These true discoveries were contrasted with features that may have been observed during this time but were not *recognized* for what they were.

The first few decades of telescopic astronomy were responsible for identifying the fundamental morphologies visible on Jupiter. Foremost among these were the intrinsic spots.

I also addressed some cosmological implications of looking at Jupiter as a world for the first time and mentioned some practical aspects of 17th-century astronomy that carry over into our present day: the quests for technological improvements and increased financial support.

Endnotes

[All titles are written in full, with the exception of '*Phil. Trans.*' for the *Philosophical Transactions of the Royal Society*.]

1. Drake S 1957 *Discoveries and Opinions of Galileo* (Garden City: Doubleday)
2. Bell L 1922 *The Telescope* (New York: McGraw-Hill)
3. Van Helden A 1974 The telescope in the seventeenth century *Isis* **65** 38
4. Pannekoek A 1961 *A History of Astronomy* (London: Allen and Unwin). Antonie Pannekoek (1873–1960); Dutch astronomer; Professor, University of Amsterdam.
5. Cocks E and Cocks J 1995 *Who's Who on the Moon: A Biographical Dictionary of Lunar Nomenclature* (Greensboro, NC: Tudor)
6. King H 1955 *The History of the Telescope* (Toronto: General Publishing). This telescope was named after Johannes Kepler of Praha.
7. Forbes G 1921 *History of Astronomy* (London: Watts). George Forbes, Professor, University of Edinburgh.
8. Webb T and Espin T 1893 *Celestial Objects for Common Telescopes* revised edn (London: Longmans Green)
9. Débarbat S and Wilson C 1989 The Galileon Satellites of Jupiter from Galileo to Cassini, Römer and Bradley *Planetary Astronomy from the Renaissance to the Rise of Astrophysics* Part A (Cambridge: Cambridge University Press) p 144

10. Chambers G 1889 *A Handbook of Descriptive and Practical Astronomy* 4th edn (Oxford: Clarendon)
11. Cocks E and Cocks J *Op. Cit.*
12. King H *Op. Cit.*
13. Van Helden A *Op. Cit.*
14. *Ibid.*
15. Bell L *Op. Cit.*
16. Pannekoek A *Op. Cit.*
17. Abetti G 1952 *The History of Astronomy* (London: Abelard-Schuman). Giorgio Abetti, Italian astronomer; Director, Arcetri Observatory.
18. King H *Op. Cit.*
19. Forbes G *Op. Cit.*
20. See, e.g., Le Monnier P 1741 *Histoire Celeste* (Paris: Royal Observatory). Pierre Le Monnier (1715–1799), French astronomer.
21. Abetti G *Op. Cit.*
22. Van Helden A *Op. Cit.*
23. Campani G 1665 An accompt [*sic*] of the improvement of optik glasses *Phil. Trans.* **1** 2
24. Pannekoek A *Op. Cit.*
25. Hooke R 1665 A spot in one of the belts on Jupiter *Phil. Trans.* **1** 3
26. Auzout A 1665 A further account, touching Signor Campani's book and performances about optik-glasses *Phil. Trans.* **1** 70 (1665)
27. Oldenburg H 1665 Some observations concerning Jupiter. Of the shadow of one of his satellites seen, by a telescope passing over the body of Jupiter *Phil. Trans.* **1** 143
28. Auzout A *Op. Cit.*
29. Campani G and Auzout A 1865 Signor Campani's answer: and Monsieur Auzout's animaduersions theron *Phil. Trans.* **1** 75
30. Oldenburg H *Op. Cit.*
31. Webb T and Espin T *Op. Cit.*
32. Holton G 1973 *Thematic Origins of Scientific Thought: Kepler to Einstein* (Cambridge: Harvard University Press)
33. Oldenburg H 1665 Of a permanent spot in Jupiter: by which is manifest conversion of Jupiter about his own axis *Phil. Trans.* **1** 143
34. Cassini G 1666 A more particular account of those observations about Jupiter, that were mentioned in numb. 8 *Phil. Trans.* **1** 173
35. Divini E 1666 Some particulars, communicated from forraign [*sic*] parts, concerning the permanent spott [*sic*] in Jupiter; and a contest between two artists about optik glasses, etc *Phil. Trans.* **1** 209
36. *Ibid.*
37. See, e.g., Dick T 1838 *Celestial Scenery; or The Wonders of the Planetary System Displayed* (Brookfield, MA: Merriam)
38. Chapman C 1968 The Discovery of Jupiter's Red Spot *Sky and Telescope* **35** 276

39. Cassini G 1672 A relation of the return of a great permanent spot in the planet Jupiter, observed by Signor Cassini, one of the Royal Parisian Academy of the Sciences *Phil. Trans.* **7** 4040

40. Chapman C *Op. Cit.*

41. Cassini G *Op. Cit.*

42. *Ibid.*

43. *Ibid.*

44. See, e.g., Gehrels T (ed) 1976 *Jupiter* (Tucson: University of Arizona Press)

45. Hooke R 1666 Some observations lately made at London concerning the Planet Jupiter *Phil. Trans.* **1** 145

46. Hooke R *Op. Cit.*

47. *Ibid.*

48. Cassini G 1686 Dècouverte d'une tache extraordinaire dans Jupiter, forte à l'Observatoire Royal, par Monsieur Cassini de l'Acad. R. des Sciences 1686. *Journal des Sçavans* 294

49. Hooke R *Op. Cit.*

50. Cassini G 1672 A relation of the return of a great permanent spot in the planet Jupiter, observed by Signor Cassini, one of the Royal Parisian Academy of the Sciences *Phil. Trans.* **7** 4039

51. *Ibid.*

52. Cassini G 1666 A more particular account of those observations about Jupiter, that were mentioned in numb. 8 *Phil. Trans.* **1** 171

53. Cassini G 1672 A relation of the return of a great permanent spot in the planet Jupiter, observed by Signor Cassini, one of the Royal Parisian Academy of the Sciences *Phil. Trans.* **7** 4040

54. *Ibid.*

55. *Ibid.*

56. Maraldi G 1708 Observations du retour de la tache ancienne de Jupiter *Histoire de l'Académie Royale des Sciences, avec les Mémoires de Mathématique et de Physique* 235

57. Cassini G 1686 Dècouverte d'une tache extraordinaire dans Jupiter, forte à l'Observatoire Royal, par Monsieur Cassini de l'Acad. R. des Sciences 1686 *Journal des Sçavans* 294

58. Cassini G 1733 Sur des nouvelles taches et des nouvelles bandes dans le disque de Jupiter *Histoire de l'Académie des Sciences, depuis son Établissement en 1666 jusqu' à son renouvellement en 1666* **2** 104

59. Cassini G 1730 Nouvelles decouvertes de diverfes périodes de mouvement dans la planéte de Jupiter, depuis le mois de Janvier 1691 jufqu'au commencement de l'année 1692 *Memoires de L'academie Royale des Sciences* **10** 1

60. Cassini G 1702 Observations de trois nouvelles taches de Jupiter *Histoire de L'academie Royale des Sciences, Année 1699* 103

61. *Ibid.*

62. Tabe I, Watanabe J and Jimbo M 1997 Discovery of a possible impact spot on Jupiter recorded in 1690 *Publications of the Astronomical Society of Japan* **49** L1

63. See, e.g., figures 3 and 4 in Cassini G 1730 Nouvelles decouvertes de diverfes périodes de mouvement dans la planéte de Jupiter, depuis le mois de Janvier 1691 jufqu'au commencement de l'année 1692 *Memoires de L'academie Royale des Sciences* **10** 1

64. Cassini J 1740 *Élémens d'Astronomie* (Paris)

65. Cassini G 1702 Observations de trois nouvelles taches de Jupiter *Histoire de L'academie Royale des Sciences, Année 1699* 103

66. Cassini G 1730 Nouvelles decouvertes de diverfes périodes de mouvement dans la planéte de Jupiter, depuis le mois de janvier 1691 jufqu'au commencement de l'année 1692 *Memoires de L'academie Royale des Sciences* **10** 1

67. Westfall R 1985 Science and patronage: Galileo and the telescope *Isis* **76** 11

68. *Ibid.*

69. *Ibid.*

70. *Ibid.*

71. *Ibid.*

Chapter 5

The Eighteenth Century

Jupiter is not a proper planet...

Sir William Herschel, Fellow, Royal Society
1781

Only a handful of papers were published on planetary astronomy in the general-interest journals of the 1700s. Indeed, regular coverage of astronomical topics had to await field-specific publications. The first of these was the *Astronomisches Jahrbuch*, begun by Johann Lambert (1728–1777) in 1774[1]. For a time, reports of jovian observations were published almost exclusively in this annual.

Much that was published during this time lies outside the scope of this book. This includes attempts to measure the diameter of Jupiter, a quantity tied up in the near-legendary 18th-century quest for the Astronomical Unit. Those who undertook to size Jupiter include Giovanni Cassini, Baron Franz von Zach (1754–1832) of Gotha and several French astronomers.

5.1   Did the Permanent Spot = the Great Red Spot?

Because the Great Red Spot appears to be the longest lived single feature on Jupiter, documentation of its early appearance is of particular significance. Thus, at this point we must consider assertions that Cassini's permanent spot was none other than the modern GRS. Morphologically, the two features are similar. The permanent spot was considerably shorter in longitude than the Great Spot, even at its modern minimum[2]; however, the widths in latitude of the two features, a more important measure because spots are constrained by the east–west jovian wind field, are equivalent.

In another argument, planetary scientist Clark Chapman notes that the phenomenon of the SEB disturbance (chapter 7) is associated with variation in the appearance of the Great Red Spot. Cassini mentioned a likely disturbance on Jupiter during a change in the visibility of the permanent spot, but the location of this activity cannot be determined from his description[3].

The rotation period for the permanent spot, as calculated by Cassini, does not match that of the Great Red Spot. Yet Chapman has reexamined Cassini's data and concluded that his assigned times were always mathematically biased by attempting to force-fit a *constant* value[4]. The apparent rotation rate of the GRS is known to fluctuate in response to the spot's own intrinsic motion. Thus, the comparison of rotation timings is unconvincing. It is unfortunate that Cassini could not leave us more precise information concerning the latitude of his spot.

The permanent spot was visible on and off to Cassini and his nephew Giacomo Maraldi (1665–1729) for 48 years. It made nine separate appearances for a total of 15 years of visibility. Chapman believes that the Great Red Spot, though considered to be in continuous presence for over one hundred years, seems to undergo periodic appearances of roughly the same frequency–if viewed through instrumentation equivalent to that available to Cassini. (Chapman made his estimation[5] based upon a numerical visibility index[6] for the GRS.)

We are left with the possibility that enough characteristics inherent in the two features are concordant–enough to grant the possibility that they are identical. It is particularly troubling, then, that, after the time of Cassini and his immediate successors, there remains a gap of over a century with no clearly identifiable sightings of this important marking. (Dessau apothecary Samuel Schwabe (1789–1875) drew a GRS-like spot on 5 September 1831[7].)

The lack of published data on the Great Red Spot through this era is inconclusive. It cannot be said whether it is the result of observational bias (as an unexpected form) or just a combination of poor resolving power, physical obscurity (or even nonexistence) and lack of serendipity. Certainly, its omission by any one individual should not be weighted too heavily.

The myth that *Robert Hooke* discovered the Great Red Spot was propagated into the 19th century by British astronomers. The first to challenge this claim was Eugène Antoniadi (1870–1944) in the 1920s. Nonetheless, books published as late as the 1990s still attributed the GRS to Hooke[8]. Marco Falorni further documents this controversy in the *Journal of the British Astronomical Association*[9].

5.2   A Jovian Mystery

Cassini first observed the permanent spot from Bologna, Italy. Bologna is also the setting for a turn-of-the-18th-century mystery concerning Jupiter.

Our mystery begins with Count Luigi Marsigli, who wished to donate his extensive natural history collection, and collection of scientific books and instruments, to the city. He hoped to establish a public astronomical observatory within an Institute of the Sciences of Bologna. However, funds were short. So Marsigli decided to make a direct appeal to the Pope[10].

But surely the pontiff routinely received such requests for municipal buildings. How to make Marsigli's stand out? Along with detailed plans for his Institute, Marsigli chose to submit a series of oil paintings illustrating the astronomical arts. He commissioned the popular artist Donato Creti to create this (successful, as it turned out) 'illustrated prospectus'[11].

Under the patronage of Marsigli, the self-taught astronomer Eustachio Manfredi (1674–1739) was already at work carrying out astronomical observations with the Count's telescopes. Manfredi, like Cassini before him, was a student of the planets. It was decided that each Creti painting would show a pleasant Italian scene, with instruments and observers in the foreground, but with a technically accurate portrait of one of the five known planets suspended in the background sky. (The remaining pictures similarly would show the Sun, Moon and a bright comet.) The idea was for Manfredi to guide the artist in producing a view of a planet as it might appear through a Marsigli telescope[12].

Here, the 1711 painting of Jupiter attracts our attention. (See the colour plate.) In it, two well dressed men observe along the banks of a dramatic cascade. One man sits with a drawing pad, while the other presumably expounds on the view of Jupiter seen through the long-focal-length refractor beside him. Above them is the telescopic view of Jupiter, incongruously large in the perspective of the picture. Fittingly, this cloudy planet nestles, joined by three of its four Galilean satellites, within a gap between dark storm clouds.

The NTB to STB appear on the disk, in order, with some shading in the polar regions. (South is at the top, as through the telescope.) The zones are white, and the belts brown. A dark spot straddles the central meridian above the STrZ. It is this spot that holds our gaze.

The spot is painted red! And herein lies our mystery. How could this be? Regardless of the identity of the permanent spot, *no* jovian feature was identified as red before the latter 19th century. Yet here it is: not in a technical drawing or observatory log description, but rather in a work of art. Was it artistic license? If so, the anachronism was the prescient guess of Raimondo Manzini, who Creti subcontracted to do the miniature work on the planetary images[13].

Or was it something more? Did Manfredi communicate to the artists a tint that he or other members of the Bologna astronomical community observed—or only suspected that they observed? Cassini was a professor at the University of Bologna until 1667. Giuseppe Campani was also from Bologna. Was the permanent spot (which reappeared in 1708[14]) literally a Great Red Spot, briefly in 1711? The Jupiter painting remains mute on this subject as it hangs today with its partners in a Vatican Museum.

5.3   The Search for the 'Surface' of Jupiter

Cassini I began a long tradition of tracking the motion of spots across the giant planet. Indeed, into the 20th century, spots were used foremost as jovian rotation period indicators. These features were scrutinized primarily so that they could be recognized repeatedly as unique markers, and then timed for multiple transits, to increase the precision of these measurements. This procedure gives insight into the perceived physical nature of Jupiter: the idea that there existed but one rotation period for the planet was an outgrowth of the belief that Jupiter, like other well-observed planets in the solar system (e.g., the Earth and the Moon), had a solid surface. If a feature were in some way connected to this 'surface', its transit time would yield the planet's rotation. Variations in rotation timings by this method often were attributed to experimental error. Cassini's determination that spots near the jovian equator travel consistently more swiftly than their poleward counterparts was stubbornly spurned.

Progress during the 18th century was slow. In Francesco Bianchini's (1662–1729) book illustrations (1737), Jupiter is depicted as seen on 6 October 1702, and in September 1713 (two views). Only the most obvious albedo features are in evidence: the NEB (1713) and SEB (1702 and 1713)[15]. (Bianchini had already calculated the (incorrect) rotation period for Venus, based on observations of nonexistent cytherean spots[16].) Yet gradually resolution improved and more features were seen.

By the late 1700s it was recognized that certain features behaved like terrestrial atmospheric phenomena (again, an Earth-based analogy) and were imbedded in global cloud patterns such as the zones. Thus, they were likely to be affected by the local windspeed prevalent at a given latitude and not representative of the motion of the 'surface' hidden below. While such a cloud pattern might be permanent, thereby obstructing a view of any truly anchored features, observers kept studying Jupiter in the hope that they might be able to glimpse a hole in the cloud deck that would reveal a 'surface'.

The dark jovian belts were watched especially because it was believed that they might represent gaps in a higher, white cloud cover and serve as windows to a lower level. Now we know that even if this is indeed the case, the belts reveal, not a solid 'surface', but merely lower, browner clouds that, of course, do not admit a single planetary rotation period for all latitudes.

Early jovian astronomers believed that rarer dark spots had more physical significance than bright ones. The latter were only of interest for rotation timing, and at that there was a strong bias toward large, well-defined, equatorial features. More diffuse or lower contrast temperate features might have been overlooked, because of technical limitations, or simply ignored.

5.4   Northward

The stage next shifts to northern Europe. This geographical transfer of prominence within planetary astronomy was a gradual one and only seems abrupt because it occurred during a natural gap in the historical record. (This gap was a 'sin of omission, not commission' and is described below.)

Technical reasons can be cited for the shift: the availability of optical glass and instrument-makers, for instance. So can political ones: new expressions of national prestige, the needs of growing maritime economies and interference from ecclesiastical authority elsewhere.

Social factors should not be disregarded, either. The relative prosperity of the North gave opportunity for endeavours not directly related to utilitarian ends. It is too easy to overlook, in the case of Britain, governmental stability and long periods of peace as parents of a conducive environment for scientific enterprise. (Naval and expeditionary wars do not count in this regard.) And then there is chance. Without William Herschel arriving on its shores, would England's observational astronomy have failed to take root, and would German achievements have reached a fuller potential?

Prior to the middle of the 19th century, the study of the visible surface† of Jupiter lagged behind that of other physical phenomena associated with the planet. While references are made to jovian surface features early in the history of the telescope, after that they are sporadic and describe only the most unusual and prominent ones.

(Jupiter was still a favourite nighttime test object for lensed telescopes. The bright disc on a black background betrayed variance from **achromaticity** with unnatural green or purple borders[17].)

During the 18th century, the observation of the physical body of Jupiter, as well as that of the other planets, was inhibited first by the reluctance of professional astronomers to use the telescope[18] and then, once this was overcome, by an accelerating trend in astronomy away from planetary observations altogether and toward stellar research, ironically now practical as a result of the advent of the large-aperture telescope. This shift was abetted by a new philosophical stance that implicitly stated that the solar system by and large was understood‡.

Most of the early published observations of the giant planet not directly involving rotation timing were simply excerpts from observing logs. Examples are those of French astronomer Charles Messier (1730–1817) who hunted for comets with a $7\frac{1}{2}$ inch Gregorian **reflector** (109×). When observing Jupiter, Messier's prime interest was in noting satellite transits and shadow crossings,

---

† The noun 'surface' is a convenient expression frequently used to refer to the thin, visible, outer layer of the jovian atmosphere. It will be used here without quotation marks but should not be misconstrued so as to imply a sharply defined boundary or phase transition. 'Surface' (in quotation marks) will continue to be used to mean an imagined '*jova firma*'.

‡ Only one sentence is devoted to specific observations of the discs of planets—for the purpose of rotation timing—in the 793 pages of Abraham Wolf's 1938 *A History of Science, Technology, and Philosophy in the Eighteenth Century* (London: George, Allen and Unwin).

but he did take time to describe briefly the appearance of the 'upper belt' and 'middle belt'[19].

A few poor drawings of Jupiter from before 1850 were published[20], but this appears to represent a feeling of obligation to 'fill in' the record more than anything else. No other explanation seems reasonable because now, as then, nothing meaningful can be extracted from them. An exception, I will show, is the work of William Herschel.

5.5   The Shape of Jupiter

The geometrician Colin MacLaurin (1698–1746†) was one of those who first undertook a mathematical analysis of the gravitation envisioned by his friend Isaac Newton. (MacLaurin was a professor at the University of Edinburgh.) This led MacLaurin to study fluid **ellipsoids** of rotation. He was highly successful and became a founding member of the Royal Society of Edinburgh.

An astronomer on the side, MacLaurin was surprised by the rapidity of variation on Jupiter—more than that seen on any other planet. He read reports by Cassini of new features that seemed to appear in an hour's time. MacLaurin incorporated this phenomenon into his main-line research and theorized that 'spring tides' caused by the jovian satellites are responsible for the upheaval evident on that planet[21].

Jupiter was, after all, an ellipsoid. MacLaurin maintained that this fact, combined with the high angular velocity of Jupiter and the 'greatness of his body', created conditions that resulted in monstrous tides[22].

MacLaurin's tidal theory seemed to work out but could not explain Cassini's observation that some spots travel at different rates than others. MacLaurin's theory would have spots travelling in the wake of the satellites and slaved to the velocities of these bodies. That was not observed to be the case. 'This is a phaenomenon (*sic*), of that kind, of which it is perhaps best not to attempt any explication, till the same be confirmed by more observations', cautioned MacLaurin[23].

Tides meant oceans to MacLaurin. He said, 'Water is of too great importance, in natural operations, to suppose hastily any planet to be deprived of it; tho' we must also allow that the variety of nature is not to be limited by our conceptions'[24]. MacLaurin encouraged astronomers to note the relative positions of the major jovian satellites when they saw a disturbance taking place or a feature forming on the disc.

These remarks, along with mathematics supporting his thesis, were published posthumously in an Edinburgh journal. They were eclipsed by the succeeding article describing one Benjamin Franklin's invention of the lightning rod. Yet, MacLaurin was to be an example to Scotch astronomers of the *19th* century who, fighting abominable observing conditions, established a tradition of important theoretical planetary work.

† A victim of the Scottish Rebellion of 1745.

Figure 5.1: Colin MacLaurin. (From a plate accompanying Wolf A 1939 *A History of Science, Technology, and Philosophy in the Eighteenth Century* (New York: MacMillan).)

5.6 The Herschels

The foremost English astronomer of the 1700s was a prolific writer, but Sir William Herschel† (1738–1822) wrote only one paper about features on Jupiter[25]. Even this paper was primarily about astronomical time-keeping.

Herschel started out by stating that the **diurnal** period of the Earth is the standard reference of time. Astronomers keep track of time in units of seconds, minutes and hours—all tied to the length of the day. Yet is the day really constant in length? What if the Earth's rotation period varies ever so slightly? Herschel worried that this could go unnoticed. For instance, if a

---

† William Herschel remains one of the pivotal figures in the history of astronomy, but most of his work was not germane to planetary astronomy. The reader is referred to *William Herschel and the Construction of the Universe* by Michael Hoskin 1963 (London: Oldbourne).

Figure 5.2: William Herschel. (Courtesy of Yerkes Observatory.)

heavenly object appeared to be slowing, was it being impeded by some sort of Cartesian medium (the popular interplanetary 'æther')? Or was it because the Earth 'clock' was accelerating and making the other object appear to slow in comparison? Now if the Earth has only a *nearly* invariable rotation period, the same also might be true for the other planets. So, to obtain a truly objective time, Herschel proposed that the rotation periods of all the planets be monitored and compared as a check on one another[26].

Herschel set about looking at existing work done on the rotation of the planets. (Indeed, it was Herschel who first estimated the period of Saturn.) He found Cassini's day-rate for Jupiter to be not accurate enough. He blamed this on the fact that jovian spots vary their countenance too often:

> Nor do the dark spots only change their place, which may be supposed to be large black congeries of vapours and clouds swimming in the atmosphere of Jupiter; but also the bright spots, though they may adhere firmly to the body of Jupiter, may undergo some apparent change of situation by being differently covered or uncovered on one side of the other, by alterations in the belts[27].

Herschel himself experimented with timing light and dark spots and obtained diverse periods[28]. He cited the possibility of equatorial winds on Jupiter that mimic those on the Earth and might impart upon different atmospheric features different velocities. Herschel presumed that such jovian winds would blow at enormous speeds compared to their Earthly counterparts.

Ultimately, Herschel preferred Mars as a more attractive celestial clock because the morphological features glimpsed on it seemed to remain constant in both shape and colour. He assumed them to be permanently fixed to the planet[29].

Herschel was a pioneer of reflecting telescopes (largest $=$ 12 inches in aperture) and used his 7, 10 and 20 foot focal-length instruments in his study of planetary rotation. In addition, he used a quadrant with telescopic sights. A pendulum kept time[30].

We would know very little about how Jupiter actually appeared in Herschel's day except for the fact that he included with his discourse on time portions of his observing log that dealt with his brief study of the planets. Thus we find Herschel in 1778 following dark spots on Jupiter[31]. Regrettably, Herschel was even farther behind continental planetary observers in planet-consistent cartography; it is impossible to tell from his notes where these spots were. (See the description of Herschel's drawings to follow.)

In 1779 Herschel recorded a bright spot in the EZ(N). Based on his analysis of the behaviour of this spot, he concluded that it and similar ones accelerated to the speed of the prevailing winds after their formation. Herschel speculated that if such an acceleration was witnessed in some seasons and not others, it would be evidence of the existence of jovian monsoons. He doubted, though, that Jupiter's small obliquity would introduce such dramatic seasonal weather patterns[32]. No one seems to have subsequently made the 12 years of observations necessary to test this hypothesis.

(William Herschel was at least partly culpable for leading observational astronomy away from the planets. On the other hand, he, almost peripherally, made an important step toward considering Jupiter in a meteorological context. Today, we know that tremendous zonal winds do blow across Jupiter. However, they are believed to be driven principally by kinetic energy transferred to them by eddies rising from the planet's deeper atmosphere—and not by solar **insolation** as Herschel implied. This is a fundamental difference between Earth weather and that in the jovian upper atmosphere.)

It is not until we encounter the work of Herschel, and the contemporaneous quality of telescope optics, that much useful information can be gleaned from artwork. Herschel recognized, and soon it became common working knowledge, that the range of magnifying powers producible with a telescope was not open ended in a realistic sense. Incredibly high ratios of objective focal length to eyepiece focal length usually had served only to magnify the effects of poor **seeing**. Magnification had to be chosen carefully for a given night (or hour) and extended object, according to what sky conditions would allow, to maximize resolution. (Herschel used 300× or less[33].) Observations and

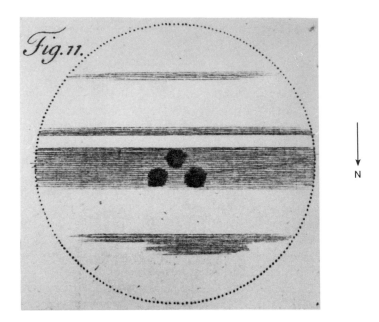

Figure 5.3: Jupiter, 7 April 1778, by W Herschel. (From table V 1781 *Philosophical Transactions of the Royal Society* **71**; courtesy of the Royal Astronomical Society.)

subsequent drawings clearly benefited from this realization.

Herschel undertook the first series of Jupiter drawings. He made his 13 illustrations for the *Philosophical Transactions*[34] on 24 February, 25 February, 2 March, 3 March, 14 March, 7 April, 12 April and 14 April 1778. The artistic style is extremely linear and impedes interpretation. It is clear that his motivation mainly was to document equatorial features suitable for rotation-timing. Still, his drawings surpass those of any predecessor.

The Jupiter of February 1778 exhibited the NTB and STB[35]. Both the NEB and SEB were 'cloudy' and indistinguishable from the NTrZ and STrZ, respectively. The EZ was relatively dark except for a large convective element in the southern component that left a white streak, partially circumnavigating the planet, in its wake. In Herschel's illustrations Nos 5 and 6 (25 February), this streak appears as a thin zone extending across the entire disc, excluding a single, dark, spot-like section. This 'spot' is larger in Herschel's illustration No 9 (3 March). Starting with Herschel's illustration No 10 (14 March), the streak appears to have caught up to its source. As its beginning was not documented, its rate of spreading cannot be calculated. (The single hemisphere views do not yield the feature's longitudinal extent at the time of the first drawing.)

In Herschel's illustrations Nos 11 and 12 (7 and 12 April) three large dark spots show up in the EZ(N)[36]. They are arranged in a triangle. While white spots are not uncommon in a clear EZ, dark spots are normally attributable to satellite shadows. The size, quantity and arrangement of these spots preclude such an interpretation.

The NTZ and the STZ appear occasionally in the drawings, and their resolution can be used as an indicator of good seeing. The night of 14 April (Herschel's illustration No 13) is an example[37]. In this drawing, the dark spots in the EZ(N) have been replaced by a white, irregularly shaped feature. The tapering of this feature to the west at its southern base is consistent with the known wind field. There is nothing else quite like it in the corpus of jovian observations. Is it an equatorial plume? (See chapter 6.)

As for Herschel's triangle of large spots, they remain among the largest 'spots' ever reported on Jupiter. However, they 'faded' in a few days[38]. Their close grouping is reminiscent of the overlapping Comet Shoemaker–Levy 9 impact sites. Yet Herschel's spots differ from one another in both longitude and *latitude*. They could not easily be created by the impact of co-orbital bodies.

In chapter 7 we will learn of phantom features called 'port holes'. Because of their placement, the equatorial spots of William Herschel can easily be interpreted as high-contrast views of this phenomenon; therefore, they are not likely to be true spots at all.

William Herschel's apparition is characterized best by the unusually extensive high albedo that occurred in the principal belts for at least six weeks. This ephemeral circumstance led him to conclude that zones do not represent the jovian 'ground'. He was the first to correctly surmise this and did so contrary to the opinions of his 17th-century predecessors.

Herschel's only son, John, a more careful draftsman, also drew Jupiter. His 23 September 1832 rendering, made with a 20 foot focal-length reflector, illustrates his *Outlines of Astronomy*[39]. At this time, Jupiter presented a more typical aspect: zones and belts alternated in albedo starting with the NTZ and continuing to the STZ. The NEB showed components, and only the NEB(S) indicated any possible partial obscuration of a belt.

5.7 Further Contributions from the Continent

The reign of the Cassinis had all but come to a close by the late 18th century, and the centre for jovian observation made the second of its moves—this time from the Paris Observatory to several smaller institutions in Germany. The rise to prominence of German astronomers, some reasons for which I suggested at the beginning of this chapter, had begun even earlier. William Herschel himself was from Hannover and only emigrated when the French occupied the ducal house.

The first major German planetary astronomer was Johann Schröter (1745–1816). His initial published examination of Jupiter took place in the winter of 1786. He was the first to write about the appearance of both the tropical

N
↑

Figure 5.4: Jupiter, 23 September 1832, by J Herschel. (From plate III accompanying Herschel J 1849 *Outlines of Astronomy* (Philadelphia: Lea and Blanchard).)

zones and the temperate belts. Later he saw additional bands[40]. Schröter noted the aspect of old and new jovian bands, and he recorded changes in their widths and albedos.

Prior to this, Schröter had studied law at Göttingen. He became a royal secretary in Hannover. A meeting there with the Herschel family revived a college curiosity about astronomy. After becoming the chief magistrate of Lilienthal in 1781, he had the time and resources to pursue his hobby. He was the first to attempt a map of the planet Mars[41].

Schröter ordered telescopes from all over Europe, including those of William Herschel. At one time, Schröter's Herschel-built 27 foot focal-length (20 inch aperture) reflector was the largest in continental Europe.

Schröter investigated the jovian rotation rate by measuring both bright and dark spots *at the same time*. He found a much greater variation in rotation period than had been reported by Cassini and Maraldi[42]. The Magistrate could think of no reason why the rotation of Jupiter as a whole should *change*[43].

Perhaps, Schröter reasoned, some of the features he was watching were really suspended at much greater elevations than others. (High clouds came to mind.) Such features still could be observed on the Earth-facing hemisphere of Jupiter when their lower counterparts had already disappeared around the

JOHANN HIERONYMUS SCHROETER
Königl. Großbrit. u. Churf. Braunschw. Lün.
Ober-Amtmann zu Lilienthal
geb. d. 30 Aug. 1745
zum Andenken gestochen von George Fischbein im Octobr. 1791

Figure 5.5: Johann Schröter. (Courtesy of William Sheehan.)

jovian limb. This would affect his rotation-time measurement. Thus, by (incorrectly) trying to force-fit a uniform rotation period for Jupiter, Schröter (correctly) concluded that Jupiter must have an atmosphere reaching great heights[44]. (See chapter 9.)

To Schröter, changes in the colours, albedos and widths of jovian bands only made sense if they were features of Jupiter's atmosphere (particularly the belts, a step backward from Herschel)[45]. He speculated that there were jovian winds blowing both eastward and westward (at different latitudes) within the belts. They carried with them spots, some pushed along with the planet's rotation and some pushed against it[46]. He also considered the possibility of winds shifting direction with the seasons: jovian monsoons[47]. Above all, he wondered why few had contested the Cassini rotation period determination (brought into question by Cassini himself) for so long.

On 7 April 1792, Schröter saw a circular spot, bright in its centre with a dusky shading around the circumference. (He had seen similar ones in 1786

N

Figure 5.6: Hubble Space Telescope image of Jupiter, showing multiple Comet Shoemaker–Levy 9 impact spots in the southern hemisphere. (Courtesy of Space Telescope Science Institute.)

and 1787[48].) The spot was in Jupiter's Southern Hemisphere. A drawing of this observation appeared (uncredited) in an 1854 edition of *Arago's Popular Lectures on Astronomy*[49]. After the modern discovery of the Great Red Spot in 1878, Schröter's observation was put forth in one of the earliest attempts to find historical precursors for the GRS[50].

In the light of Comet Shoemaker–Levy 9's spectacular July 1994 collision with Jupiter, William Sheehan and Thomas Dobbins speculate that Schröter might have observed something similar[51]. They quote Schröter's observation of a *pair* of colatitudinal spots, both extremely dark and well defined. While SL-9 did fragment before impact, and therefore produced a series of colatitudinal dark spots on Jupiter, there is no good way to distinguish Schröter's spots from intrinsic ones—based only on his brief description.

5.8   Why Observe Jupiter?

I have written of the sparsity of jovian observations from this period. It is time to examine 'the other side of the coin'. Why was Jupiter as attractive an object for study as it *was*?

Practically, as a bright and extended object in the sky (one of the few appropriate for the crude optics and pointing accuracy of the 17th and 18th centuries), it ranks fourth after the Sun, Moon, and Venus. It does not create the technical difficulties that come with observing the Sun. Except near conjunction, its whole disc is continuously visible; it does not wax and wane.

From a philosophical point of view, not only the markings on Jupiter but also their changing nature were a direct challenge to an Aristotelian picture of the Universe. 'Enfin la variabilité de ce monde est telle', said a later student of Jupiter, 'qu'il offre à l'observateur et au penseur un des plus nouveaux et des plus intéressants problèmes de l'astronomie planétaire' [At last the variability of this world is such ... that it offers the observer and the philosopher one of the most recent and interesting problems in planetary astronomy].[52]

Observations of satellites and planetary rotation did not provide the sole evidence for Copernicanism in the jovian system during the Enlightenment. I would submit that while positional measurements of the satellites—bright objects against the black background of the sky—were easier to make, subtler physical observations of Jupiter itself posed an equal number of questions, the answers to which pointed in the direction of the modern concept of the solar system.

5.9 Summary

In the age of Newton, Jupiter lost some of its metaphysical interest. Modern astronomy began to move northward to protestant countries, where it flourished. Technical reasons such as the availability of quality glass and engineering skills abetted this transition (ironically, to locations of exceptionally poor astronomical viewing conditions). Now, however, situated in maritime nations such as England, telescopic astronomy was transformed into a positional (i.e., navigational) science. When it resumed its role as a form of exploration of the physical environment, largely at the hands of William Herschel, it was with the intent of comprehending a universe of stars, not planets. Thus, as the technology that would improve the capability of planetary astronomy evolved, that sub-field left the mainstream of astronomy. For a time during the 18th century only a few individuals, such as Schröter, continued to be interested in both documenting phenomena on Jupiter and attempting to interpret them.

Yet, the positive legacy from the 1600s and 1700s, benefiting the study of Jupiter, comes out of the connection between astronomical and nautical science existing during this time—not so much in the realm of observation, but in interpretation. As European sailors roamed the terrestrial seas, astronomers who stayed at home began to look upon Jupiter as a fluid world. This was a viewpoint that would become enormously useful in trying to understand the visible disc of the planet.

Endnotes

[All titles are written in full, with the exception of '*Phil. Trans.*' for the *Philosophical Transactions of the Royal Society*.]

1. Herrmann D 1984 *The History of Astronomy from Herschel to Hertz-sprung* (Cambridge: Cambridge University Press)
2. Chapman C 1968 The Discovery of Jupiter's Great Red Spot *Sky and Telescope* **35** 276
3. *Ibid.*
4. *Ibid.*
5. *Ibid.*
6. Defined by Peek B 1958 *The Planet Jupiter* (London: Faber and Faber)
7. *Ibid.* Schwabe had discovered the eccentricity of Saturn's rings four years earlier.
8. See, e.g., Leverington D 1995 *A History of Astronomy from 1890 to the Present* (London: Springer); Lankford J (ed) 1997 *History of Astronomy: An Encyclopedia* (New York: Garland); and Shirley J and Fairbridge R (ed) 1997 *Encyclopedia of the Planetary Sciences* (London: Chapman and Hall)
9. Falorni M 1987 The discovery of the Great Red Spot of Jupiter *Journal of the British Astronomical Association* **97** 215
10. Bedini S 1980 The Vatican's astronomical paintings and the Institute of the Sciences of Bologna *Proceedings of the Eleventh Lunar and Planetary Conference* **1** xiii
11. *Ibid.*
12. *Ibid.*
13. *Ibid.*
14. Maraldi G 1708 Observations du retour de la tache ancienne de Jupiter *Histoire de l'Académie Royale des Sciences, avec les Mémoires de Mathématique et de Physique* 235
15. Bianchini F 1737 *Veronensis Astronmicae ac Geographicae Observationes Selectae* (Verona: Manfredi)
16. Cocks E and Cocks J 1995 *Who's Who on the Moon: A Biographical Dictionary of Lunar Nomenclature* (Greensboro, NC: Tudor)
17. See, e.g., Dollond G, Herschel J and Pearson W 1826 Report of the Committee appointed by the Council of the Astronomical Society of London, for the purpose of examining the telescope constructed by Mr Tulley, by order of the Council *Monthly Notices of the Royal Astronomical Society* **2** 507
18. Van Helden A 1974 The telescope in the seventeenth century *Isis* **65** 38
19. Messier M 1769 A series of astronomical observations made at the Observatory of the Marine at Paris *Phil. Trans.* **59** 457
20. See, e.g., Dick T 1838 *Celestial Scenery; or the Wonders of the Planetary Solar System Displayed* (Saint Louis: Edwards and Bushnell). I own a copy of this book and have found a number of inaccuracies in the text.

21. MacLaurin C 1754 Concerning sudden and surprising changes observed in the surface of Jupiter's body *Essays and Observations, Physical and Literary, Read Before a Society in Edinburgh* **1** 184
22. *Ibid.*
23. *Ibid.*
24. *Ibid.*
25. Herschel W 1781 Astronomical observations on the rotation of the planets around their axis made with a view to determine whether the Earth's diurnal motion is perfectly equable *Phil. Trans.* **71** 115
26. *Ibid.*
27. *Ibid.*
28. *Ibid.*
29. *Ibid.*
30. *Ibid.*
31. *Ibid.*
32. *Ibid.*
33. *Ibid.*
34. *Ibid.*
35. *Ibid.*
36. *Ibid.*
37. *Ibid.*
38. *Ibid.*
39. Herschel J 1849 *Outlines of Astronomy* (Philadelphia, PA: Lea and Blanchard)
40. Schröter J 1788 Sur la rotation et l'atmosphere de Jupiter *Journal de Physique ou Observations et Memoirs sur la Physique, sur D'histoire Naturelle et sur les Arts et Metiers* **32** 108
41. Cocks E and Cocks J *Op. Cit.*
42. *Ibid.*
43. Schröter J 1783 Allgemeine Folgerungen und Bemerkungen über die Rotation und Atmosphäre des Jupiters *Beilragezuden Neusten Astronomischen Entdeckungen* 104
44. *Ibid.*
45. *Ibid.*
46. Schröter J 1788 Sur la rotation et l'atmosphere de Jupiter *Journal de Physique ou Observations et Memoirs sur la Physique, sur D'histoire Naturelle et sur les Arts et Metiers* **32** 108
47. Schröter J 1783 Allgemeine Folgerungen und Bemerkungen über die Rotation und Atmosphäre des Jupiters *Beitragezuden Neusten Astronomischen Entdeckungen* 104
48. Johnson S 1880 Spot on Jupiter in 1792 *Observatory* **3** 283
49. Dennett F 1881 Jupiter in 1792 *Observatory* **4** 58
50. Johnson S *Op. Cit.*
51. Sheehan W 1996 Personal communication. See also, O'Meara S 1996 Schröter and Jupiter's dark spots *Sky and Telescope* **91** 98
52. Flammarion C 1877 *Terres du Ciel* (Paris: Librairie Académique Didier)

Chapter 6

The First Half of the Nineteenth Century

Next to Pallas†, in order of the system, is the planet Jupiter.... When nearest to the earth, at the time of its opposition to the sun, it is about 400,000,000 of miles distant from us. A faint idea of this distance might be acquired by considering that a cannon-ball flying five hundred miles every hour, would require more than ninety-one years to pass over this space; and a steam-carriage, moving at a rate of twenty miles an hour, would require nearly 2,300 years before it could reach the orbit of Jupiter.

Thomas Dick, LLD
1838

A casual reader of Dieter Hermann's revisionist (marxist) history of astronomy[1] might conclude that nineteenth-century astronomy was synonymous with German astronomy. Insofar as Jupiter was concerned, this was nearly so through the 1840s.

6.1 Observations from Germany ('Barges')

At the beginning of the new century, 'Hofrath' Huth of Frankfurt noticed three things about the jovian disc: first, the north polar region appeared darker than the south; second, the northern hemisphere seemed 'flatter' than the southern; and third, the bands of Jupiter were slightly convex to the North[2]. From these astute observations, he concluded that Jupiter's rotation axis was not perpendicular to the planet's orbital plane. Jupiter exhibited an obliquity—although a small one. Huth does not seem to have been aware of Cassini's much earlier work on this subject.

† Pallas is the most distant from the Sun of the four largest asteroids within the Asteroid Belt.

57

Not every recorded discovery about Jupiter was real. In 1821 an avocational astronomer from Poland, Justice Commissar Georg Kunowski (1786–1846), was excited by reports that Privy Counselor Johann Pastorff (1767–1838) of Buchholz had found 'Photosphären' surrounding Venus and Jupiter[3]. (Pastorff more frequently observed the Sun.) This phenomenon consisted of luminous spheres some distance from the planetary discs. In the case of Jupiter, the Galilean satellites orbited within the radius of illumination.

Kunowski, observing with an exquisite $4\frac{1}{3}$ inch aperture Fraunhofer refractor, confirmed Jupiter's 'Photosphäre' and measured it to be 25 arc minutes in diameter. He also discovered an identical sphere surrounding Saturn—of the same angular extent! His suspicions aroused, Kunowski pointed his telescope at the bright star Sirius—and, later, Vega—and, under good seeing conditions, identified 25 arc minute spheres around each of them[4].

Reluctant to give up such a tantalizing idea as planetary 'Photosphären', Kunowski nonetheless devised an experiment that was to prove their nonexistence[5]. He returned to Jupiter and observed it as it set behind the roof of a neighbour's house. If the luminous sphere were real, it would start to be **occulted** *before* the planet but would continue to be visible just *after* the jovian disc was blocked from sight. Instead, the sphere was visible in its entirety until the moment Jupiter disappeared, at which time it vanished.

Kunowski was forced to conclude that the 'Photosphären' were optical effects produced by the reflection of bright objects on an interior surface within a compound lens. The commissar used a 6 foot focal-length telescope. As Pastorff's spheres were consistently 16 arc minutes in extent, Kunowski wagered that he used a shorter focal-length refractor. A Dollond† telescope obtained by Kunowski showed the same effect exactly[6].

The polymath Karl Gauss (1777–1855) (who considered Pastorff a dilettante) told the counsellor to cover a portion of his objective and see what happened to the photospheres. Pastorff was the only one left unconvinced by the result[7].

With a literary sigh, Kunowski marked the passing of planetary photospheres: 'So ist denn die physische Astronomie um eine Entdeckung ärmer, und die Optic hat ein kleines Problem gewonnen' [This is how physics in astronomy is poorer of a discovery and optics has gained a small problem][8].

(Pastorff would not be the last to confuse optical effects with atmosphere. Using a newly borrowed telescope, Edouard Gand would see a 'Nébulosité diaphane, comparable à un immense globe de cristal, au centre duquel brillent les cinq astres du monde jovien [a translucent nebulosity, comparable to a huge crystal globe, at the centre of which the five stars of the jovian world shine], [Jupiter plus four satellites] nearly 60 years later[9].)

---

† John Dollond (1706–1761) perfected an achromatic lens in 1757. Eventually, refracting telescopes so equipped would show 'the disc of Jupiter as white and as free from colour as a reflector' (telescope collector William Kitcher quoted by King H 1955 *The History of the Telescope* (Toronto: General Publishing).)

Figure 6.1: Johann Mädler. (Courtesy of Patrick Moore.)

A mid-19th-century German astronomer was Friedrich Bessel (1784–1846), Johann Schröter's one-time assistant†. The first person to determine the distance to a star also kept track of jovian spots[10]. Bessel's accurate calculation of the density of Jupiter yielded a result much lower than that for the Earth.

Another was Johann Mädler (1794–1874). Like many astronomers, Mädler turned to the planets after seeing a comet in his youth. While much of his professional work dealt with improving rotation-period determinations, Mädler also wrote about how Jupiter looked. Many of his papers were co-authored by a patron, the banker Wilhelm Beer (1797–1850), who had purchased Pastorff's telscopes and established a private observatory[11].

Mädler's first contribution was a small note in November 1834, mentioning that 'auf der Jupitersscheibe ausser andern Flecken, zwei sehr schwarze, und scharf begränzte in sehr geringer Entfernung von der nördlichen Aequatorzone zeigen' [shown on the Jupiter disc amongst other spots are two very large, black and sharply bounded, an insignificant distance from the northern equatorial zone] [12]. (Here we see for the first time the use of the word 'Aequatorzone' for the EZ—though the NEB may be implied.) Later a third

† Both Bessel and Schröter used telescopes manufactured by William Herschel.

spot was seen[13], but it, the authors concluded, may simply have been a dark part of the band.

The three features followed each other at intervals of 40 minutes. Of the two well defined spots, the following one increased in size as the astronomers tracked it. They did so through April 1835[14]. Three spots, simultaneously visible for so long a period, afforded an excellent opportunity for establishing an even-more-precise rotation time. This Mädler and Beer set out to do, believing it to be the first attempt since Cassini[15].

During these months we also have their description of the whole jovian disc. In December, for instance, they saw two bands (presumed to be the NEB and SEB) that were unequal. The northern belt was much fainter and narrower than the southern[16]. Mädler and Beer had a **micrometer** that they used together with their $3\frac{3}{4}$ inch aperture Fraunhofer refractor to make their historic map of the Moon in 1836. They were less successful in their attempts to measure the jovian belts and publish numerical results.

Germany has no better weather than the rest of northern Europe. During January 1835, Mädler and Beer experienced 17 consecutive nights of clouds. When Jupiter was again visible to them, the NEB had completely disappeared. Fortunately, the spots that they were tracking remained. Meanwhile, the SEB had increased in both width and darkness[17]. Thus, the German astronomers inadvertently documented a significant short-term change in the appearance of the planet.

Unbeknownst to Mädler and Beer, Professor George Airy (1801–1892) also was observing Jupiter at this time, just before being named Astronomer Royal in England. He wrote of 'a remarkable spot seen on the apparent southern belt, nearly four times as large as the shadow of the first satellite, very well defined. About two-thirds of its breadth was apparently below the belt, and one-third upon the belt'[18]. On 13 December, there were two spots 'on the apparent lower belt, both well defined. The following spot is larger'[19].

Airy's description of these spots, seen through the Cambridge refractor (aperture $= 11\frac{3}{4}$ inches; focal length $= 19$ feet[20]), matches that of Mädler and Beer in terms of arrangement. (The following was larger.) Airy contributed further valuable information on the morphology and dimensions of the spots—he compared the latter to the shadow cast by Io—that is missing from the report of Mädler and Beer[21]. He also provided more detail on the location of the spots. However, here the descriptions do not coincide. Mädler and Beer wrote of a northern feature, while Airy called it a southern one.

Mädler and Beer published a list of transit times, and so did Airy. Scanning these tables, one sees that three dates appear on both. That is, both sets of astronomers were able to obtain a transit time on the same day. On 23 December 1834, all three astronomers watched the requisite transits, though Mädler and Beer did so just after midnight and Airy just before. On 6 January 1835, Airy recorded a transit at 2 h 22 m 22 s[22]; Beer and Mädler observed the same transit at 3 h 20 m 38 s and 3 h 21 m 20 s, respectively[23].

(The deviation between the latter two times reflects the differing personal

Figure 6.2: George Airy. (Courtesy of Yerkes Observatory.)

equations of these observers. A 'personal equation' is introduced in the reduction of astronomical measurements that require assigning precise absolute times to events observed by eye. It is a way of quantifying the reaction time of the observer and, if present, his or her tendency to record certain types of event early or late. It usually takes the form of a single term that is the result of testing the observer under artificial conditions where the true time of the 'event' is known.)

Both parties recorded local sidereal time[24], Airy at Cambridge and Beer and Mädler at Beer's private observatory in Berlin. The discrepancy of one hour is commensurate with the difference in longitude between the cities. The eastern observatory reported a later sidereal time. From all of this, we can conclude that, for the first time, there were **astrometric** timings from independent observers of a single event, each of which corroborated the other. (Jean Valz (1787–1861) observed the same spots from France, but not on the same dates as Mädler and Beer or Airy[25].)

On 16 January, Airy's time was 1 h 55 m 23 s[26] for a spot transit; that same night one observer in Germany (probably Mädler) obtained 2 h 8 m 2 s[27]. There is a greater difference here, at least some of which is perhaps accountable for by the differing observing procedures used by the three astronomers. Mädler and Beer attempted to estimate the passage of the centre of the spot across the central meridian, while Airy concerned himself with watching the edge contacts. (One can measure the transit time of an extended feature by estimating the location of the middle of the feature and then recording the time at which this point crosses the appropriate meridian; alternately, one can record the times at which both the leading and the trailing edges of the feature contact the meridian and then average the two.)

The conflicting accounts of the location of the spot also can be reconciled. Airy used the key phrase 'apparent southern belt'[28]. For positional astronomers such as Airy, this referred to the belt that appeared at the bottom of a telescope's field of view. Inverted, this would be a northern feature. From Mädler and Beer we can infer that the spot was at the arbitrary boundary between the EZ and the NEB. (Recall the description of dark belts.) Airy's description would put it closer to the NTrZ†.

Unfortunately, the only other published drawings from this period are two included in a book entitled *Celestial Scenery*[29]. They purport to picture Jupiter as it appeared in 1832–33 and in 1837. Both illustrations are confusing. No sense can be made of the seemingly random assortment of bands that cross the jovian disc in these prints. There are no recognizable spots.

The huge 'well-bounded' Mädler and Beer/Airy spots were used as examples of canonical jovian spots in several later textbooks[30]. These illustrations show them as even larger features than their written descriptions suggest. Indeed, they seem to nearly rival the famous dark spots produced on Jupiter by the Comet Shoemaker–Levy 9 impacts in 1994[31].

The actual described location, darkness, and multiplicity of the Airy/ Mädler and Beer features suggest a different origin for these spots. They might be extremely large examples of the morphological form known as 'barges'. The cyclonic 'barges' are found at +14° latitude. They translate eastward at only a few metres per second. (The rotation time for the Airy/Mädler and Beer spots was shorter than Cassini's, once both had been corrected for light-travel time.)

'Barges' are easy to see because of their contrast with their surroundings. They would have been particularly interesting to early 19th-century observers who would have equated them with great depth. Spotting such a feature would have brought the observer closer to seeing the 'real' Jupiter, so it was thought.

Mädler left Berlin in 1840 for the Dorpat Observatory. After his departure, Beer appears to have undertaken no more serious astronomical or topographical work.

(As a postscript to the list of influential German Jupiter-watchers should be added the name of selenographer Johann Schmidt (1825–1884), though he became Director of the Athens Observatory. His attempts to remeasure the planet's rotation period spanned from the 1860s[32], to the 1870s[33], to the modern appearance of the Great Red Spot (1878)[34].)

Airy's rotation period for Jupiter (9 h 55 min 21 s) became so accepted that, when the rotation periods of two more-equatorward spots were mea-

† Hans-Jörg Mettig calculates that the rotation periods of the spots of Mädler and Beer are more consistent with the windspeed in the NTrZ than that in the EZ.

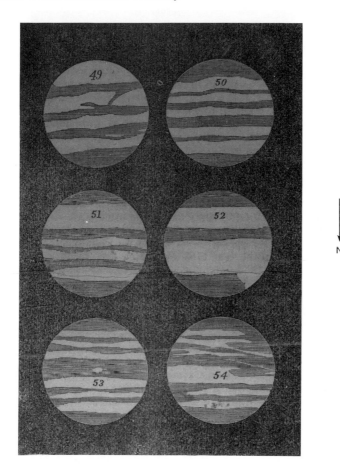

Figure 6.3: Jupiter, 1832–1833 and 1837, by Dick. (From Dick T 1838 *Celestial Scenery; or the Wonders of the Planetary Solar System Displayed* (Saint Louis: Edwards and Bushnell) Reprinted in 1854, p 77.)

sured in 1876, Sir George's period was still the standard reference. The faster rotation period measured for the 1876 spots was solely attributed to proper motion of the spots themselves; the bulk of the planet must behave as Airy said it would![35]

Interpreting features on Jupiter as phenomena in a visible planetary atmosphere was a positive step. However, it brought with it a tendency to make a one-to-one correspondence between meteorological activity on Jupiter and that on the Earth. That extrapolation of Earth atmospheric physics to systems that were themselves the size of the Earth *itself*—or nearly so—might be an error of scale was not a major concern at the time and still hinders some students of the giant planets.

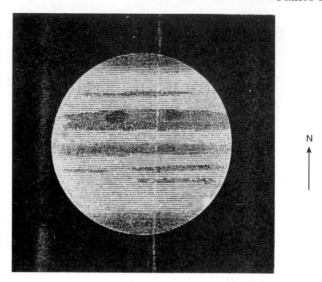

N
↑

Figure 6.4: Jupiter, *circa* 1840, by Mädler. (From a plate accompanying Beer W and Mädler J 1841 *Beitrage zur Physischen Kenntniss der Himmlischen Körper im Sonnensystem* (Weimar).)

6.2   The State of Knowledge at Mid-19th Century

We have now followed planetary astronomy through a time when its pursuit was clearly secondary to that of stellar astronomy in the eyes of professional astronomers. We have seen nearly as many different styles and techniques of observing as reasons for doing so in our study of Jupiter-watching from northern Europe: from William Herschel's peripheral but detailed descriptions of Jupiter, to Schröter's, Beer's and Mädler's searches for the giant planet's rotation-time, to dead-ended investigations of imagined planetary phenomena. Real newly discovered jovian features observed during this time were low-albedo ones, typified by the northern hemisphere 'barges'.

In contrast, the second half of the 19th century was a high point in the history of planetary astronomy. More than 200 observatories were established during the 1800s, most after 1850, and the instruments inside them were markedly improved over those of a century earlier. An ensemble of people was beginning to come together with a common interest in observing the planets and in developing a common technical language with which they could communicate with each other via the new vehicle of journals. These individuals shared a consensus on a philosophy governing the procedure of their inquiry. Called Neo-Baconianism, it was never acted upon in a pure and unmodified way, but offered a near-universal starting point for entering into the still infant science.

While this period was to be a fruitful one, it began with a humble set of accepted facts about the major planet Jupiter. An anonymous review article in the *Sidereal Messenger* provides such new information as that the belts of Jupiter appear 'ragged and torn' and that the belts and spots appear to *change*, and sometimes disappear completely, over a period of as little as a few months[36]. Otherwise, everything in the 1847 article easily could have been written a century earlier.

Two years later, Sir John Herschel (1792–1871) described Jupiter succinctly (yet to the extent of knowledge available at the time of publication) in his extremely popular textbook, *Outlines of Astronomy*:

> The disc of Jupiter is always observed to be crossed in one certain direction by dark bands or belts . . . These belts are, however, by no means alike at all times; they vary in breadth and in situation on the disc (though never in their general direction). They have even been seen broken up and distributed over the whole face of the planet; but this phenomenon is extremely rare. Branches running out from them, and subdivisions... as well as evident dark spots, are by no means uncommon; and from these, attentively watched, it is concluded that this planet revolves... on an axis perpendicular to the direction of the belts. Now, it is very remarkable, and forms a most satisfactory comment on the reasoning by which the spheroidal figure of the Earth has been deduced from its diurnal rotation, that the outline of Jupiter's disc is evidently not circular, but elliptic, being considerably flattened in the direction of its axis of rotation. This appearance is no optical illusion, but is authenticated by micrometrical measures, which assign 107 to 100 for the proportion of the equatorial and polar diameters. And to confirm, in the strongest manner, the truth of those principles on which our former conclusions have been founded, and fully to authorize their extension to this remote system, it appears, on calculation, that this is really the degree of oblateness which corresponds, on those principles, to the dimensions of Jupiter, and to the time of his rotation[37].

This paucity of information was about to diminish. The 1850s brought with them the first of a steady stream of observational papers on the *appearance* of Jupiter by English-speaking astronomers that continues unabated today. (This venerable observing tradition has been continued unabated right up to the present day by the British Astronomical Association, which routinely publishes amateur drawings of Jupiter alongside *in situ* video images and high-resolution CCD images.) This marked the change in planetary work from the domination of German to British and then American astronomers spoken about in the first chapter. It also was the start of a jovian record rich enough—that is, with short enough periods between observations—to enable the dynamics of the jovian disc to become truly apparent.

Perhaps what is most important, it is midway through the last century that efforts to 'measure' Jupiter (e.g., rotation-timings), an act almost demanded by the natural history of the time that emphasized the determination of numbers—recall John Herschel's invoking of 'micrometrical measures' in the passage quoted above, began to give way to qualitative description. The quantitative style of science had plateaued regarding Jupiter, given the limitation of 'grounded' technology and lack of theoretical understanding. The qualitative style was in no way similarly fettered and benefited from technical achievements in optics. Communications about Jupiter now centred on the form and substance of what was seen there, not the geometrical parameters of the planet. Jovian astrometry would remain as a parallel programme of inquiry, at least in continental Europe. However, it would not attract the vital body of amateurs who would soon take on the major burden of planetary science involving Jupiter. Jovian descriptive astronomy would not reach *its* technical limit for several decades more. As an example of this divergence, compare the aforesaid work of Schröter and Mädler to that of the individual I now introduce.

The first paper on jovian surface morphology published in the United States (1848) was also the last on the subject by a pioneer of American astronomy, William Bond (1789–1859). (Subsequent papers list the name of Bond's third son, George Bond (1825–1865), as primary author; indeed, during the latter period of the elder Bond's life, the two worked together so closely that it is difficult to separate their contributions.) William Bond, like so many other students of Jupiter, started out as an amateur astronomer. Unlike many others, he was poor. Bond learned about scientific instruments working in his father's clock shop as a youth. There, he also became interested in astronomy and eventually constructed his own telescopes. He seems to have spent much of his income on this hobby. In 1839, Harvard College invited him to move his private observatory to its grounds, and Bond became the first (unsalaried) Director of the Harvard College Observatory.

The senior Bond's paper[38] is a straightforward account of the appearance of the jovian disc on two dates: 28 January and 3 February 1848. Bond described nine 'belts' including the Polar Regions (seen through a 15 inch aperture refractor[39]). One interpretation of this is that he may be speaking of light bands (zones) rather than dark belts. (Bond went on to compare the 'belts' to cirrus clouds, which are white.) If this is so, the inversion of what was to become the standard nomenclature showed up again when Bond spoke of 'the principal equatorial belt', which is then the Equatorial Zone[40]. (He noted the nearly parallel edges of this band.) It is confusing, however, that he also described the next most northerly 'belt' (zone) as irregular—particularly on its northern edge. Today the north tropical zone is normally irregular on its *southern* edge. It could be the case that the NTrZ appeared then merged with a clouded-over north temperate belt (NTB) and was indistinguishable as a separate zone. Yet this is unlikely if Bond saw nine distinct light regions because the planet's belts would have had to have been clear and dark to separate the light bands.

Figure 6.5: William Bond. (Courtesy of Lick Observatory.)

An alternate interpretation would assume that Bond used the word 'belt' for any band on Jupiter. Then, a subset of the NTZ, NTB, NTrZ, NEB, EZ, SEB, STrZ, STB, STZ and Polar Regions would comprise nine 'belts'. This would mean that the irregular belt was the NEB.

However, Bond elsewhere described intervals *between* the 'belts' as 'curd-ling'[41]. This sounds more like a description of modern nomenclature belts and is evidence favouring the first interpretation.

Bond could distinguish only three 'belts' in his observations of the other hemisphere six nights later. He described the 'broad' band (the EZ) again and said that it straddled a line 'a little south' of the equator[42]. He also noted, as a change from previous observations, that its edges were no longer parallel. If, during the apparition, the SEB was 'clouded-in', the EZ and SEB could be counted as one undifferentiable and very wide 'belt'. This would partially explain the paucity of 'belts' seen by Bond but would cause the mid-line of the combined region to appear much more than 'a little south' of the equator. It would be especially true if the southern edge of the white SEB blended into the STrZ and none of these bands, the EZ, SEB or STrZ, were distinguishable from one another. On the other hand, if Jupiter was so well banded less than a week before, it is much easier to blame this word picture on poor seeing.

On the same night, a deep hollow indented the SEB's southern edge. Today we are tempted to interpret this as an early observation of the Great Red Spot Hollow. If it is, then the famous Great Spot itself may have been undergoing a low-contrast phase of its existence. It seems that Bond would certainly have noted even the faint appearance of this remarkable feature unless he was totally colour insensitive. Instead, Bond mentioned the indentation in passing and went on to discuss what he considered more important phenomena: the broken appearance of the 'principal' northern 'belt' (perhaps a 'clearing' NEB) and some dark spots within it[43]. These spots could be identical with the NEB 'barges' of today.

The region known as the Great Red Spot Hollow (which often has a red spot situated within it) is an anticyclonic vortex that translates at a few metres per second in longitude. As it does, it appreciably deflects the flow in the SEB, causing a characteristic 'dent' in the southern edge of that belt. This indentation shows up even *without* the presence of the distinctive Red Spot.

Periodically, there is none of the characteristic red chromophor visible in the Hollow at all[44]. Why is this?

Normally, the SEB is dark and the Great Red Spot leaves what looks like a wake to the west of itself. This manifestation is easily recognized by the contrast in colour and brightness. At other times, though, there is a great deal of high-albedo material in the belt. As eddies of this material flow by the GRS, they may be deflected, split up or caught by the vortex. These phenomena, difficult to observe from the Earth, are evident in Voyager and Galileo images. The latter is a mechanism for 'contaminating' the GRS chromophore. At times, ammonia ice might completely conceal this agent. (The supply of available ammonia ice is regulated by conditions in the adjacent belt.) During these times, the feature in question only can be detected morphologically as the Hollow. Even in high-resolution photographs, it is difficult to differentiate it then from its surroundings in any other way. Such may have been the case in Bond's time.

William Bond's article was considered significant enough to be reprinted in Europe[45]. An edited version appeared in *Astronomische Nachrichten* without translation (and without credit to the *Proceedings of the American Academy of Arts and Sciences*, in which it originally appeared).

This attention to such limited observations highlights the difficultly in obtaining good views of Jupiter in northern climates. (Jupiter never rises to a high altitude in the sky at these latitudes and thus must always be observed through much atmosphere; see below.) Bond himself cited the need for a 'tranquil atmosphere' for viewing[46]. A combination of favourable weather and atmospheric seeing conditions was all too rare for European and North American astronomers of the day, and reports of observations, made under conditions when the theoretical resolution of the telescopes in use could be approached, were doted upon.

6.3 Recurring Spots in the South Temperate Zone

Another observer struggling to monitor Jupiter through uncooperative skies was William Lassell (1799–1880), an English brewer-turned-astronomer. Lassell is famous for his discovery/confirmation of satellites of Neptune (Triton in 1846), Saturn (Hyperion in 1848) and Uranus (Ariel and Umbriel in 1851), but in between, on 27 March 1850, he made another sort of discovery[47]. Using his 24 inch aperture Newtonian† (focal length = 20 feet, magnification = 430×) in good seeing conditions, Lassell discovered a series of bright white spots in the south temperate latitudes of Jupiter unlike any that had been seen before‡. As the planet rotated, these spots kept their spatial relation with respect to one another, thereby proving that they were features intrinsic to the planet.

Lassell sketched what he saw and presented the drawings to the Royal Astronomical Society (RAS)[48]. The belt/zone pattern is skewed if reckoned from the +24° jet. This suggests that Lassell began drawing in the north, starting with the jet, and proceeded southward. In the southern hemisphere he put in fine detail, including a broad STB and a prominent but dim STZ. It was this hemisphere in which he was most interested. Lassell's spots form a zigzag pattern across the STZ. These distinctive white spots are identical with those seen in Voyager images in 1979 and Hubble Space Telescope images of the Shoemaker–Levy 9 impact sites in 1994. Only their size is exaggerated. Thus, far from being an occasional phenomenon, Lassell's spots were a manifestation of a potentially frequently occurring jovian feature.

Lassell's paper is a useful example of how astronomical discoveries of the day were communicated among astronomers. Word of Lassell's spots spread quickly by 19th-century standards. This was due, in part, to the recent appearance of journals devoted exclusively to astronomy. The transition from letters as a means of conveying scientific information to journals is reflected in the propagation of Lassell's announcement: a report of Lassell's description of the white spots to the RAS appeared in the *Monthly Notices of the Royal Astronomical Society* along with a woodblock print of one of his drawings[49]. This report reappeared shortly thereafter in German translation. It was part of a letter to the editor of the *Astronomische Nachrichten*, Heinrich Schumacher (1780–1850), from an English correspondent of his, The Reverend Richard Sheepshanks (1794–1855)[50]. (Sheepshanks was himself editor of the *Monthly Notices of the Royal Astronomical Society*[51].) It then made its way across the Atlantic as a letter from Schumacher to Benjamin Gould (1824–1896), who

† A 'side view' reflecting telescope in which a secondary mirror moves the focus off the objective mirror's optical axis.

‡ Lassell's spots were the first important jovian phenomenon to be reported in the new (1847), expanded *Monthly Notices of the Royal Astronomical Society* (See Dreyer J and Turner H 1923 *History of the Royal Astronomical Society: 1820–1920* (London: Royal Astronomical Society).)

Figure 6.6: William Lassell. (Courtesy of the Royal Astronomical Society.)

published it in the first volume of his upstart *Astronomical Journal*[52]. There it appears—in English, again—as Schumacher quoting Sheepshanks quoting Lassell!

Observations of white spots of the kind described by Lassell were not reported again until over seven years later. A country doctor and one-time minister named William Dawes (1799–1868) had observed features in the Spring of 1849 that he later decided were identical with Lassell's spots, but 'being otherwise much engaged, did not observe them closely'[53]. (Dawes was in chronic ill health much of his life.) He did not see these features again distinctly until September 1857. In his paper, he described the reappearance of the spots and included sketches.

Using an 8 inch aperture refracting telescope, Dawes counted a total of five spots with two being equal in size and almost as large as the disc of Ganymede. These spots travelled across the planet in close proximity to each other. West of them, three more spots existed, which Dawes described as smaller than the disc of Europa. The spots, except for the middle of the five, formed a co-latitudinal line. (Dawes was the first to record numerical latitudes for jovian features.) Two days later Dawes could detect slight positional

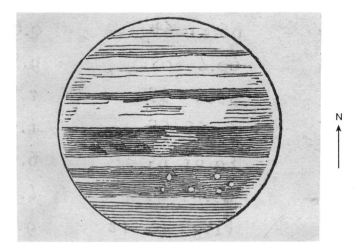

N

Figure 6.7: Jupiter, 27 March 1850, by Lassell. (From 1850 *Monthly Notices of the Royal Astronomical Society* **10** p 134; courtesy of the Royal Astronomical Society.)

changes among the spots. By 28 October, they were spread out over a greater range of longitudes, and five new spots seemed to join them. On 30 October, another new spot appeared among the set. Dawes concluded that the one white spot that lay south of the others was translating eastward faster than its companions[54]. This is consistent with Voyager spacecraft velocity-versus-latitude measurements for the vicinity of the STZ.

By November, the belt that the white spots inhabited had become white except for a few dark streaks. However, Dawes now reported that approximately five light spots had appeared in the belt equatorward of the STB (the SEB). Dawes also noted that the STB and SEB were tenuously separated during this time by a disjoint STrZ[55].

Dawes continued his commentary on the white spots later in the year[56] and produced more sketches. He described their changing relative positions and the continued partial separation of the STB and SEB by a semiobliterated STrZ. He also reported that William Smyth (1788–1865), a retired Admiral in Her Majesty's Navy and now a magistrate and amateur astronomer[57], had also observed the spots.

In his third note[58], Dawes said that the spots in the 'second southernly belt [STB and STZ]' still may be seen. He then digressed to recite an all-too-familiar story of the detection of a dark spot, which through ephemeris confusion turned out to be the shadow of Ganymede. Dawes used this accident, however, to make a spatial comparison: the diameter of the largest STZ spot was two-thirds that of the shadow of Ganymede. In a sentence about his later observations of the white spots (in December 1857), Dawes used the word 'oval' to describe two of them.

Figure 6.8: William Dawes. (Courtesy of the Royal Astronomical Society.)

William Dawes' initial artistic treatment of Lassell's spots consists of four drawings from the autumn of 1857[59]. As many as eleven STZ spots appear at one time in these drawings.

Dawes' second series of 1857 drawings exhibits more features[60]. In the south, the SEB and STrZ are divided in the right portion of the disc by the GRS Hollow. No longer a mere indentation or stylized oval, it appears here in its modern 'pinched' ellipsoidal shape. There is a classic pileup of white material in the SEB immediately behind the GRS Hollow. The StrZ remains brilliantly white. The STB and STZ are present as a series of lines that continue into the SPR.

One drawing[61] (Dawes' illustration No 1) is reminiscent of a higher- resolution version of John Herschel's[62]. The general effect of the picture is such that it could well be mistaken for a crude facsimile of a 20th-century spacecraft image.

The next night[63] (Dawes illustrations Nos 2 and 3) the GRS Hollow had rotated off the disc (though its imminent emergence may be indicated by a widening of the SEB at the east limb). Some white clouds remain in the SEB at the latitude of the northern edge of the GRS Hollow. There are a few spots

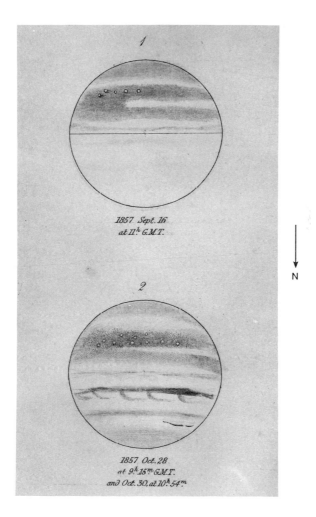

N

Figure 6.9: Jupiter, 16 September 1857 and 28 October 1857, by Dawes. (From a plate accompanying 1858 *Monthly Notices of the Royal Astronomy Society* **43**; courtesy of the Royal Astronomy Society.)

in the STZ. Dawes' illustration No 4 (5 December) shows the GRS Hollow partially rotated off the disc.

Now William Lassell re-entered the story of the spots to which he originally brought attention in 1850. In a notice to the Royal Astronomical Society, he wrote '... I cannot deny myself the pleasure of sending you a drawing of the planet *Jupiter*, as seen last night with this instrument—or rather early this morning'[64].

Lassell[65] noted that, coincident with the period in which white spots were not documented, Jupiter was at its southernmost **declination**. It is clear

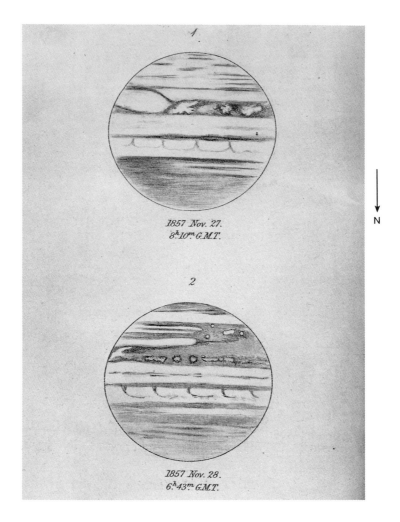

N

Figure 6.10: Jupiter, 27 November 1857 and 28 November 1857, by Dawes. (From a plate accompanying 1858 *Monthly Notices of the Royal Astronomy Society* **43**; courtesy of the Royal Astronomy Society.)

that Lassell believed these spots to have been in continuous existence since at least 1850, though not observable because of the unfavourable position of Jupiter in the sky, and not to have been in any way discovered or rediscovered by Dawes.

After some general comments about the overall appearance of Jupiter (the belts were narrower and lighter than usual), he went on to describe, in addition to his spots, a set of *new* bright spots in the EZ. These he compared to the original STZ spots as more 'delicate'[66].

Lassell described two dark oblong features equatorward of the Lassell spots visible for two months late in 1858[67]. He suggested that they were the beginning of the restoration of the SEB, which had not been distinguished easily in recent apparitions. The white spots were translating westward more rapidly than these features. (German artist Herrmann Goldschmidt (1802–1866) saw a single dark feature that looked as if it were separating in two[68].)

Lassell's drawings of the STZ spots made one year later than Dawes' (Lassell's illustration No 1, 18 November 1858, and Lassell's illustration No 2, 5 December 1858) lack the detail of the Reverend's[69]. However, the first figure shows a distinctive characteristic of the spots: dark annulae surrounding them. However, such annulae may be simply a draftsman's convention, used to accentuate the brilliance of the spots.

The final observations of these STZ spots recorded in the literature[70] occurred in September 1859, and were made by Sir William Murray (1801–1861) with a 9 inch aperture Cooke† refractor[71]. He said that, at that time, two or more of the spots were nearly co-longitudinal. He also drew attention to the fact that the 'principal belt [EZ] had quite a flocculent cloudy appearance' and presented 'a very mottled aspect'[72].

William Smyth, as part of his *Cycle of Celestial Objects*[73], summarized the history of the white spots that same year. He recorded the aforementioned observations and added to the list those made by Professor Schumacher in Altona.

Smyth gave Dawes precedence on the issue of discovery, the rule of first *publication* not having been firmly established at that time. Also, by this stage in his career, 'eagle eyed'[74] Dawes had done significant work in the measurement of double stars with a micrometer and just lost out on first discovering Saturn's Crepe Ring by ten days (to the Bonds). Dawes had even reported the presence of a (dark) spot on Jupiter in 1843[75]. In 1860 his reputation probably approached that of Lassell, though it was the satellite-finder who was to achieve more lasting fame.

Smyth departed from strictly observational reporting and suggested a physical cause for the spots. Citing their static relative positions (in fact, they were not always so), he asked whether the spots might be (borrowing Murray's words) 'ice and snow seen through openings in their clouds?'[76] This false (but self-consistent) interpretation again illustrates the eagerness of astronomers of the era to fit Jupiter into the familiar planetary model.

The next significant outbreak of spot reports would not occur until 1870.

In modern times, small, bright, anticyclonic spots are found at approximately $-41°$ latitude. They are slightly oval in shape. Voyager images have shown that many are surrounded by lower-albedo annulae; some show a degree of spiral structure within their interiors.

The STZ/Lassell spots 'zigzag' across the planetary disc—each succeeding spot in longitude is alternately a little north or south of the mean latitude. It

† Murray, Dawes, and many other English observers used telescopes with lenses made by master craftsman Thomas Cooke (1807–1868).

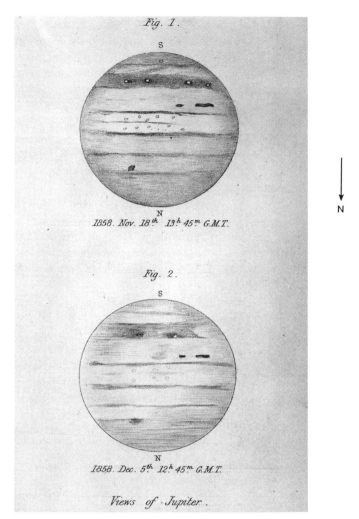

N

Figure 6.11: Jupiter, 18 November 1858 and 5 December 1858, by Lassell. (From a plate accompanying 1859 *Monthly Notices of the Royal Astronomical Society* **19**; courtesy of the Royal Astronomical Society.)

is this facet that is exaggerated in the early drawings. They also are separated by dark cyclonic cells, seeming to highlight the white spots, that have a 'folded filamentary morphology'[77] and often prevent the STZ from looking like a traditional 'zone' at all.

The spot density as a function of longitude during the 1850s is made difficult to assess by the loss of contrast that affects any feature more than 45° east or west of the central meridian on the jovian disc. While it appears that observers noticed a single group of spots moving across the planet, it is unclear whether, once the location of this group was specified, similar (perhaps smaller or lower-contrast) spots were sought after at other longitudes. Today, when they are apparent, STZ spots extend all the way around the planet; in the past, even small variations in atmospheric transparency and steadiness with time may have made them seem to be longitudinally dependent.

(While there is no evidence for cyclical appearances of Lassell's spots as was suggested by some 19th-century observers, their resolvability would vary through the course of the jovian year. In reality, all claims to seasonal behaviour must be weighted against the beat of good seeing 'windows' that observers at high latitudes experience once every twelve years. I will discuss this phenomenon more fully in a later chapter.)

6.4   Equatorial Plumes

William Dawes also described a jovian phenomenon that will become important to our discussion later. While monitoring the white spots, he documented the sighting of 'festoon-shaped shadings' south of the NEB[78]. These, in turn, reminded him of features he had seen and sketched in 1851. They were 'five large and nearly regular arches' in the EZ(S) which indented the SEB[79]. Dawes had not seen them before or since. However, Bond's[80] uneven equatorial zone could have been a poor-resolution version of these shapes.

Lassell[81] and Murray[82] independently reported 'streaks' north of the SEB. Today, arcuate cloud structures in the EZ(N) are called 'plumes'. In isolation, their curved nature may not be noteworthy, and they appear as 'streaks'. At other times, they may be arranged periodically around the planet in such a way as to seem to come in pairs—therefore, perhaps, the 'arch' resemblance.

EZ(N) plumes form when material rising from a convective region—the head of the plume—spreads out in the wind shear present in the equatorial zone. They originate at +10° latitude and travel at 100 m s^{-1} eastward. Plumes resemble terrestrial cumulus clouds. In fact, EZ(N) plumes are similar to the 'anvil head' clouds seen in the Earth's atmosphere. The analogy may be extendable to include '...violent turbulent motions, very high precipitation rates, hail production, lightning, and thunder'[83].

Twelve to 15 plumes may gird the planet at any one time. They are regularly spaced, suggesting a wave effect, but the shapes of individual plumes can differ. Occasionally, one or more plumes will lack the characteristic 'head', but most do possess it. This feature, along with the length of the plume (which may effectively take up the entire latitudinal extent between the northern edge of the Equatorial Zone and the equator), makes them easy to see.

A factor contributing to the variety in their descriptions is the extreme variability of the plumes. They can change dramatically in less than one diurnal rotation, making keeping track of individual ones difficult. The 'festoons' Dawes observed in 1857 eventually encircled the planet and evidently were a very dramatic apparition of the phenomenon.

Finally, before leaving the work of William Dawes, there is one more transient jovian feature that he does not allude to in writing but nevertheless appears in his artwork. In Dawes's first drawing of Jupiter, designed to show the STZ spots, the STrZ extends across only two-thirds of the planet's disc[84]. The abrupt albedo change at these latitudes suggests that a south tropical zone disturbance is taking place there. (During a STrZ disturbance, the zone becomes dusky within a discrete segment.) The STrZ is complete as drawn a few weeks later, but the reappearance of the possible disturbance in a still later drawing makes it likely that this longitudinally dependent darkening simply was not on the visible disc each time Dawes observed and drew the STrZ[85].

The limitation in the work of observers such as the Bonds, Lassell and Dawes was their insistence upon concentrating on individual features or groups of features that they considered new morphologies, to the exclusion of what was happening elsewhere on the planet. Thus, some events nearly were missed. We cannot be sure that their work truly reflects all of what would be of interest to us today that could be seen on Jupiter during the observational careers of these individuals.

This 'tunnel vision' approach was demanded by practical constraints. While diligent, few of these men could be called full-time professional observers. There is no evidence that they had quite the indefatigable obsession of, say, a William Herschel, who prized his time at the telescope above food and sleep. Adding to this the matter of the weather and the fact that Jupiter was not always the only planet in the sky, it becomes clear that selectivity was inevitable.

Even so, isolation and identification of temporal features on Jupiter, and the documentation of their behaviour, was a first step in decoupling the interpretation of jovian phenomena from that of those seen on the Earth. The numerous observers of Jupiter who followed Lassell, Dawes and Angelo Secchi (see below) set out to compile a more exhaustive objective record. As science became more and more specialized, those who observed the giant planet (coming as they did mainly from the ranks of amateurs) increasingly were less knowledgeable in general natural history. Perhaps they were less bound to its way of defining all of nature from the point of view of the Earth, as well. These observers usually were not interested themselves in the 'big picture' of Jupiter. (I will discuss amateurs and theory in chapter 9.) Yet without these transitional figures, it would not have been possible to evolve a truly planetary science.

6.5   A Widely Distributed Painting of Jupiter

During the 1856 apparition, Warren De La Rue (1815–1889) created a drawing of Jupiter of such artistry and (apparent) attention to detail that it was used as the closest thing to a standard visual reference in written discussions of the planet's features[86] until the end of the century (though it was never called

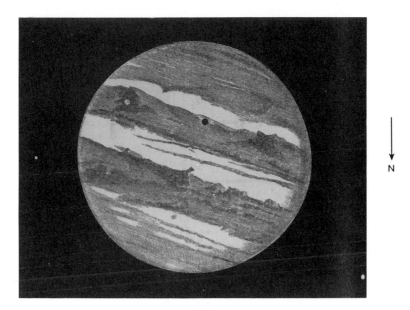

N

Figure 6.12: Jupiter, 25 October 1856, by De La Rue. (From a plate accompanying Guillemin A 1883 *The Heavens: an Illustrated Handbook of Popular Astronomy* 9th edn (New York: MacMillan).)

that). In this respect, it partly fulfilled the function that the standard Voyager **cylindrical-projection** maps do today.

This was possible because De La Rue transferred his drawing to an engraved plate. It could easily be published, and it was, widely and repeatedly. By this process it was implicitly set up as a 'good example'. Others trying to make similar drawings easily could have been influenced by it. (The same argument may be applicable to other only slightly less well reproduced jovian drawings, particularly those of famous observers.)

Yet the subtle effect of such an influence would be difficult to detect, and there is no clear evidence in subsequently published jovian pictures of intentional or unintentional copying. Perhaps most Jupiter observers, at least consciously, adopted an attitude based on the maxim that they could 'do it better themselves'. By contrast, in the post-Voyager era, ubiquitously published high-resolution images of Jupiter do seem to affect the work of artists attempting to render the giant planet on a pad via telescope and eye.

This is not to say that earlier observers did not influence each other. The medium of influence was not generally the appearance of reproduced work in (usually) popular literature, though. Instead it was the frequent one-on-one contact between amateur astronomers who often had common business, political and social interests beyond astronomy that, in addition to it, drew them together. One might be told what to expect to see at the eyepiece of another's telescope. Then (after looking for one's self, or perhaps not)

Figure 6.13: Warren De La Rue. (Courtesy of the Royal Astronomical Society.)

one might take that mentally interpreted image home (often literally just down the road) so that it might 'reappear' in the field of view of one's own instrument.

At any rate, it seemed that De La Rue was ideally suited to his task, which was to construct at least the most *popular* image from the mid-19th century. He was an exceptional technical drawer and had experience both in astronomy and printing. The son of a printer, De La Rue was one of the first to use electrotyping and coinvented the first envelope-making machine. He was introduced to the sky by another practical man and amateur astronomer, James Nasmyth. (See chapter 9.) De La Rue's telescope was a self-made 13 inch aperture reflector†, of 10 foot focal length.

Woodcut prints of De La Rue's Jupiter plate circulated. These could be obtained by any Fellow of the RAS who applied to the Assistant Secretary:

† Lassell, Nasmyth and De la Rue were unusual in that they had the time, money and skill to cast and tend to a large reflecting telescope. Maintaining these instruments involved frequent repolishing and refiguring of the speculum metal mirror.

'Fellows of the Society will do well to send for their copies in as many cases as possible, as they are of a size to be very liable to injury if forwarded through the post'[87].

Actually, De La Rue presented four drawings to the Royal Astronomical Society. Three of these were rendered by himself and one '... by my chemical assistant and friend Dr. Müller, who is a rapid and accurate draughtsman, and whose faithfulness in planetary delineation I have had many opportunities of testing'[88].

In his accompanying paper[89], De La Rue described the 'southern belt [SEB]' as consistently darker than the 'northern belt [NEB]'. The latter was normally more diffuse, being broken by intervals and streaks, but also was the broader of the two. Both belts were brown. The poles were considered yellow—more so in the North. De La Rue mentioned faint yellow streaks appearing under the 'northern belt', presumably in the EZ(N).

De La Rue made measurements to draw the jovian bands properly convex. (He found the accurate shape of the projected planetary ellipse by similar direct measurements.) Previously, belts and zones always had been rendered as straight lines, and the small but discernible inclination of Jupiter's axis had not been taken into account[90].

While the artistry of De La Rue's paintings of Jupiter has been long praised, today the works themselves yield disappointingly little information. His famous depiction of 25 October 1856 shows the equatorial zone as disturbingly narrow[91]. Incongruously wide equatorial belts border it.

Although resolution allowed visibility of the equatorial band, De La Rue pictured no belts or zones beyond the NTB and STrZ. A curved portion of the STrZ cannot be interpreted reliably as the Great Red Spot Hollow because nearly everything in De La Rue's work is curved!

The painting definitely conveys a sense of flow—that is, material flowing in the jovian wind field, but the sense of this flow is wrong. This suggests structure as much from memory and imagination as from the eyepiece. Alternately, but less likely, De La Rue made use of additional optics, not unlike the *camera obscura*, to project the jovian image before drawing, and a left–right inversion occurred. However, in his description of his drawing technique, De La Rue makes no mention of such a device.

A less well known De La Rue painting from the same year adds an STB and shows the equatorial zone and belts in equal proportion[92]. NEB and SEB components can be seen. The STB appears disturbed (unresolved Lassell's spots?). This drawing is quite possibly a higher resolution and more faithful rendering than its celebrity counterpart. Still, though, the direction implied by the contours of shapes is opposite of that predicted from the known wind field. (The ambiguity of the date leaves open the possibility that this drawing was made early in the year and represents the 1855 apparition.)

A sketch that illustrates William Smyth's *Cycle of Celestial Objects*[93], the first astronomical catalogue written expressly for amateur astronomers[94], is probably a better paragon for the period. It was made by Murray on 6 October

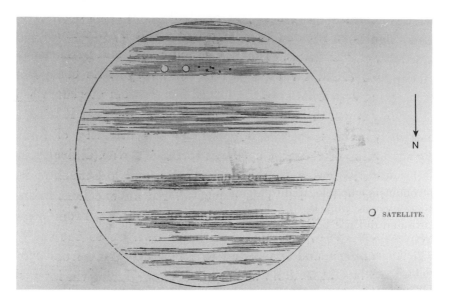

Figure 6.14: Jupiter, 6 October 1857, by Murray. (From Smyth W 1860 *The Cycle of Celestial Objects Continued at the Hartwell Observatory to 1859* (London: Nichols) p 73.)

1857 (240×–350×). Murray patiently pencilled in the belts and zones from the SSTZ to an unusual NNNTZ. There are two STZ spots.

De La Rue published portraits of other solar-system bodies. He also was the first to take collodion photographs of the Moon. While others may have tried[95], no one could surpass at least the beauty of his planetary images.

I make the concluding statement of the last paragraph with qualification. In any art history of jovian drawings, a hindrance exists. It is that not only did each artist render the same 'model' in the same 'seating' with the same 'profile' against the same background, but very little change of 'expression' was even to be expected. Adding to this the lack of the resolution available to portrait artists, whose subject is situated at a somewhat more convenient distance, few parameters remain to allow us to judge artistic style. We are left with a rather truncated interpretation, which indicates some variation of style among drawers (and some consistency of style within the figurative portfolio of a single artist in the few cases where a lengthy time span of work is available for our inspection), but nothing that gives us much further insight.

Some Jupiter-watchers drew better than others. The purpose for which a drawing was made varied as did the actual technique, but little more can be said that does not hinge on some guessed degree of fidelity or on aesthetics. I suggest the latter to be the principal reason, along with his willingness to publish, for De La Rue's work being singled out from the body of contemporary drawings for passage down to succeeding generations of astronomers.

6.6   Where To Observe?

Before the mid-1800s, astronomical observatories were located according to convenience. Most astronomers housed their telescopes near their homes. Those living in cities occasionally moved to more rural locations to avoid the first scourges of light pollution and what must have been, in northern climates, noticeable loss of transparency throughout much of the year due to coal smoke. In Britain, rural villages tended to be situated near water and consequently offered even more inclement weather, fog, and damp cold. Observing conditions there could scarcely have been better.

Astronomers who relocated to hospitable climes praised the improvements of their telescopes' abilities. Lassell used his 48 inch aperture telescope only on Malta and did not bother with more than 24 inches in aperture while in England. (And even then '. . . a favourable state of atmosphere for the use of the 20-foot equatoreal telescope is in this locality almost as rare as the visit of a comet. . . '[96].)

The miserable astronomical conditions in northern Europe were a bane to astronomers and partially account for the fact that even after William Herschel, faint-object astronomy was slow to catch on there. Jupiter was an easily observed object even through a murky sky.

It is not surprising that astronomers such as John Herschel were impressed by more southerly observing sites. (Herschel visited South Africa.) Not only was the climate generally better in these places, but more of the celestial sphere could be seen. Still, further ways of improving observing conditions were sought.

One way was to move up. It was suggested that observatories be placed at great height. A moderate increase in altitude would put the observer above significant fractions of the atmosphere and thereby enable her or him to forego looking through this obscuring medium.

That was the theory. C Piazzi Smyth (1819–1900) sought to test it in practice. Piazzi Smyth was named for the astronomer Giuseppi Piazzi by his father, William Smyth. Coming from such an astronomical tradition, the younger Smyth naturally took up the vocation. He was an assistant at the Royal Observatory, Cape of Good Hope, by the age of 26; in that capacity he learned what a telescope could do out from under Scottish skies. Named Astronomer Royal for Scotland in 1845, the transition to the Edinburgh Observatory must have reinforced in Smyth the need to set up telescopes in places favourable to their use.

When the British Admiralty funded an expedition to make astronomical observations from a high place, Smyth led it. In the 1850s the summits of great mountains were far from easily accessed. The site chosen was the 3000 m Peak of Tenerife in the Canary Islands. Piazzi Smyth took with him a $7\frac{1}{4}$ inch aperture refractor and the '5 foot equatorial presented some years since by the late Rev. R. Sheepshanks' who had bequeathed the instrument to the Edinburgh Observatory[97].

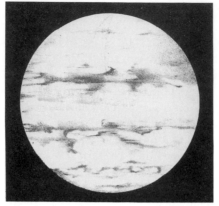

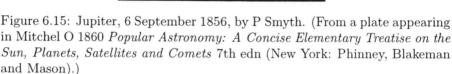

Figure 6.15: Jupiter, 6 September 1856, by P Smyth. (From a plate appearing in Mitchel O 1860 *Popular Astronomy: A Concise Elementary Treatise on the Sun, Planets, Satellites and Comets* 7th edn (New York: Phinney, Blakeman and Mason).)

Smyth was principally interested in making stellar observations for testing sky transparency and resolution. However, for checking the effect of local seeing conditions on extended objects, he made eyepiece drawings of the Moon and planets, most notably Jupiter. (In one of them, a diagonal splitting of the SEB, from the meridian to the east limb, may represent half of an enlarged GRS Hollow as it rotates onto the disc.)

Smyth was an advocate of carefully executed drawing as an astronomical tool:

> If astronomical drawing is to take a similar trustworthy and trusted place to numerical observation, in its own branch of subjects, we must in the first place with every man's work, eliminate errors in drawing and imperfections of the means and medium employed. How easily much of this may often be accomplished by *two* drawings, may be well seen... is there some doubtful mark in one? We have only to look at the other; and if the mark was in the scene itself in nature, it will likewise appear in the second view as well... [98]

That Smyth found *consecutive* nights when Jupiter could be rendered adequately at 350 power (and therefore could present all longitudes on the surface of the planet) was evidence enough of the advisability of locating observatories atop mountains. The quest for low **air mass** was to continue. Smyth's favourable report was to influence—indirectly—the placement of the new professional observatories that were to be established later in the century.

Interestingly, two of these observatories, Lick and Lowell, which were built on mountains of the American West, were to become involved with monitoring Jupiter during the 20th century. The Observatory of Pic-du-Midi de Bigorre, one of the first places where astrophysical techniques were applied to the solar system, is at an elevation of more than 2800 m.

The Keck Observatory and NASA Infrared Telescope Facility, both on Mauna Kea, Hawaii, are the best examples of modern, large, well placed reflectors used in the service of planetary astronomy. But today, the ultimate mountaintop is space, and the Hubble Space Telescope routinely produces images of Jupiter (and other planets) comparable to spaceprobe imagery.

6.7   A Latter-Day Italian Observer

Italian planetary astronomy, transplanted to an academically freer France in the 18th century, declined quickly. It staged a brief resurgence, 200 years after Cassini I, in the person of Angelo Secchi (1818–1878), himself an emigrant (temporarily) after expulsion from Rome in 1848 for activities associated with his Jesuit order. Fickle Italian politics allowed Secchi's return to his native country in 1849, and in that year he became director of the observatory at the Collegio Romano. There, in between better known astrophysical pursuits, Secchi observed Jupiter with a 10 inch aperture Merz† (focal length = 109 inches) and made micrometer measurements of the planet's disc. He stressed the significance of the equatorial *asymmetry* of the jovian bands and looked for a fundamental clue to the nature of Jupiter in it[00].

(The history of astronomy better remembers Secchi for his initial work in stellar spectroscopy. In 1863, he examined the spectrum of Jupiter. There Secchi found wide, dark absorption bands (especially at long wavelengths). He concluded that Jupiter's atmosphere contains elements different from those in the Earth's[100].)

1856 is the first instance where drawings made by multiple observers at different telescopes, but at roughly the same time, are available for our inspection. Secchi published two drawings that depict Jupiter on 10 October 1856 and 6 December 1857[101]. The former was made a little over one month after one of Piazzi Smyth's. The features tentatively identified in Smyth's image are lost; however, in the following year's Secchi drawing, the Great Red Spot Hollow is probably between the STrZ and SEB at the east limb. There also is a spot in the STZ. (Secchi was expelled again in 1873!)

6.8   Summary

A modest increase in interest in physical observations of the planets—particularly Jupiter—occurred 150 years ago. Good descriptions of both the 'normal' appearance of the giant planet and the evolution of time-dependent features became regular fare in the journals. *Bright* spots were documented. A jovian

† German optician Georg Merz (1793–1867) made lenses for Dawes, Airy and William Bond as well.

art emerged. Even the Great Red Spot Hollow appeared in drawings. Still, observers were limited, by technology and the independent nature of their work, in what they could achieve.

Endnotes

[All titles are written in full, with the exception of '*Phil. Trans.*' for the *Philosophical Transactions of the Royal Society* and '*Mon. Not.*' for the *Monthly Notices of the Royal Astronomical Society.*]

1. Hermann D 1984 *The History of Astronomy from Herschel to Hertzsprung* (Cambridge: Cambridge University Press)
2. Huth 1804 Einige astronomische-physische Beobachtungen, vom Hrn. Hofrath Huth zu Frankfurt a. d. Oder, angestellt vom Jan. bis May 1804 *Astronomische Jahrbuch* 185
3. Kunowsky J 1822 Enige physische Beobachtungen des Mondes, des Saturns, Jupiter und Mars, der Doppelsterne etc., mit einem 6-füfsigen Frauenhoferschen Fernrohr, 4-1/3 zoll Oeffnung, vom Hrn. Justiz-Commissarius Kunowsky hieselbst mitgetheilt *Astronomische Jahrbuch* 214
4. *Ibid.*
5. *Ibid.*
6. *Ibid.*
7. Schmidt T 1997 personal communication
8. Kunowsky, J. *Op. Cit.*
9. Gand E 1879 Étude sur la planéte Jupiter *Les Mondes, Revue Hebdomadaire des Sciences et de Leurs Applications* **50** 95
10. Bessel F 1840 Beobachtungen zweier Flecken auf der Scheibe des Jupiter *Astronomische Beobachtungen auf der Königlichen Universitäts-Sternwarte in Königsberg* **20** 77
11. Wolf R 1890 *Handbuch der Astronomie* (Zurich)
12. Mädler J 1835 Beobachtungen zweier Flecken auf der Jupiterscheibe 1834 November *Astronomische Nachrichten* **12** 135
13. Beer W and Mädler J 1835 Über die Rotation des Jupiter *Astronomische Nachrichten* **12** 257
14. *Ibid.*
15. Beer W 1835 *Astronomische Nachrichten* **12** 199
16. Beer W and Mädler J *Op. Cit.*
17. *Ibid.*
18. Airy G 1835 Observations of a spot on Jupiter's disk 1834 *Astronomical Observations Made at the Observatory of Cambridge by George Airy* **7** 188
19. *Ibid.*
20. King H 1955 *The History of the Telescope* (Toronto: General Publishing)
21. Airy G *Op. Cit.*
22. *Ibid.*

23. Beer W and Mädler J *Op. Cit.*
24. Airy G *Op. Cit.*; Beer W and Mädler J *Op. Cit.*
25. Valz J 1834 Auszug aus emem Briefe des Herrn Valz au den Herausgeber *Astronomische Nachrichten* **12** 239
26. Airy G *Op. Cit.*
27. Beer W and Mädler J *Op. Cit.*
28. Airy G *Op. Cit.*
29. Dick T 1838 *Celestial Scenery; or the Wonders of the Planetary Solar System Displayed* (Saint Louis: Edwards and Bushnell) Reprinted in 1854
30. See, e.g., Tuxen J 1872 *Stierneuerdenen* 3rd edn (Rjøbenhaun: Philipsens)
31. See Hockey T 1994 The Shocmaker–Levy Spots on Jupiter: Their Place in History *Earth, Moon and Planets* **66** 1
32. Schmidt J 1865 Über die Bewcgung Dunkler und heller Flecken auf Jupiter *Astronomische Nachrichten* **70** 81 and Schmidt J 1867 Beobachtungen auf der Sternwarte zu Athen im Jahre 1866 *Astronomische Nachrichten* **68** 289
33. Schmidt J 1874 Über die Rotation des Jupiter *Astronomische Nachrichten* **92** 71
34. Schmidt J 1880 Üeber den rothen Streifen auf Jupiter, 1879 *Astronomische Nachrichten* **97** 67
35. Brett J 1876 On the proper motion of bright spots on Jupiter *Mon. Not.* **36** 355
36. 1847 Jupiter and His Moons *The Sidereal Messenger* **1** 73
37. Herschel J 1849 *Outlines of Astronomy* (Philadelphia: Lea and Blanchard)
38. Bond W 1848 *Proceedings of the American Academy of Arts and Sciences* **1** 325
39. King H *Op. Cit.*
40. Bond W *Op. Cit.*
41. *Ibid.*
42. *Ibid.*
43. *Ibid.*
44. See Peek B 1958 *The Planet Jupiter* (London: Faber and Faber)
45. Bond G and Bond W 1850 Observations on the belts and satellites of Jupiter and on certain nebulae *Astronomische Nachrichten* **30** 93
46. Bond W *Op. Cit.*
47. Lassell W 1850 Sketch of Jupiter as seen on the 27th March, 1850, at 11 h 11 m G.M.T. *Mon. Not.* **10** 134
48. Lassell W *Op. Cit.*
49. *Ibid.*
50. Sheepshanks R 1850 Aszüge aus Herrn Sheepshanks' Briefen an den herausgeben *Astronomische Nachrichten* **30** 319 Neither man started out as an astronomer: Schumacher was a law student who turned to science under the inspiration of Gauss; Sheepshanks collected scientific instruments.

51. Dreyer J and Turner H 1923 *History of the Royal Astronomical Society:*
    1820–1920 (London: Royal Astronomical Society) John Dreyer (1852–
    1926), Director of the Armagh Observatory, served as President of the
    Royal Astronomical Society. Herbert Turner (1861–1930) was an Oxford
    professor of astronomy.
52. Schumacher H 1851 From Letters of Professor Schumacher to the Editor
    Astronomical Journal **1** 76. Gould was also a student of Gauss.
53. Dawes W 1858 On the appearance of round bright spots on one of the
    belts of Jupiter *Mon. Not.* **18** 6. Dawes was by now in charge of the
    private observatory of George Bishop.
54. *Ibid.*
55. *Ibid.*
56. Dawes W 1858 Further observations of the round bright spots on one of
    the belts of Jupiter *Mon. Not.* **18** 49
57. Dreyer J and Turner H *Op. Cit.*
58. Dawes W 1858 Miscellaneous Notes. *Mon. Not.* **18** 72
59. Dawes W 1858 On the appearance of round bright spots on one of the
    belts of Jupiter *Mon. Not.* **18** 6
60. Dawes W 1858 Further observations of the round bright spots on one of
    the belts of Jupiter *Mon. Not.* **18** 49
61. *Ibid.*
62. Herschel J *Op. Cit.*
63. Dawes W *Op. Cit.*
64. Lassell J 1859 Physical observations of Jupiter *Mon. Not.* **19** 51
65. *Ibid.*
66. *Ibid.*
67. *Ibid.*
68. Goldschmidt H 1859 Tache noire allangée, sur Jupiter *Cosmos, Revue*
    Encyclopedique Hebdomadaire des Progres des Sciences **14** 148
69. Lassell W *Op. Cit.*
70. Murray W 1859 Physical observations of Jupiter *Mon. Not.* **19** 51
71. 1862 William Keith Murray *Mon. Not.* **22** 108
72. Murray W 1860 Physical observations of Jupiter *Mon. Not.* **20** 58
73. Smyth W 1860 *The Cycle of Celestial Objects Continued at the Hartwell*
    Observatory to 1859 (London: Nichols)
74. Moore P 1997 William Rutter Dawes *Encyclopedia of Planetary Sciences*
    ed J Shirley and R Fairbridge (London: Chapman and Hall)
75. Chambers G 1877 *A Handbook of Descriptive Astronomy* 3rd edn (Ox-
    ford: Clarendon)
76. Smyth W *Op. Cit.*
77. Smith B *et al* 1979 The Galilean satellites and Jupiter: Voyager 2 imaging
    science results *Science* **206** 927
78. Dawes W 1858 On the appearance of round bright spots on one of the
    belts of Jupiter *Mon. Not.* **18** 6
79. *Ibid.*

80. Bond W *Op. Cit.*
81. Lassell W *Op. Cit.*
82. Murray W *Op. Cit.*
83. West R, Strobel D, and Tomasko M 1986 Clouds, aerosols, and photo-chemistry in the jovian atmosphere *Icarus* **65** 161
84. Dawes W 1858 On the Appearance of Round Bright Spots on One of the Belts of Jupiter *Mon. Not.* **18** 6
85. Dawes W 1858 *Op. Cit.*; Dawes W 1858 Further observations of the round bright spots on one of the belts of Jupiter *Mon. Not.* **18** 49
86. See, e.g., Browning J 1870 On a change in the colour of the equatorial belt of Jupiter *Mon. Not.* **30** 39; Webb T 1870 The Planet Jupiter, 1869–1870 *Popular Science Review* **9** 127; Guillemin A, Proctor R and Lockyer J (eds) 1883 *The Heavens; An Illustrated Handbook of Popular Astronomy* (New York: MacMillan); Proctor R and Ranyard A 1892 *Old and New Astronomy* (London: Longmans, Green)
87. 1857 *Mon. Not.* **17** 221
88. De La Rue W 1857 Observations of Jupiter, During October 1856 *Mon. Not.* **17** 6
89. *Ibid.*
90. *Ibid.*
91. Guillemin A, Proctor R and Lockyer J (eds) *Op. Cit.*
92. Proctor R and Ranyard A *Op. Cit.*
93. Smyth W *Op. Cit.*
94. King H 1955 *The History of the Telescope* (Toronto: General Publishing)
95. See, e.g., Winlock J 1876 *Astronomical Engravings of the Moon, Planets, etc.; Prepared at the Astronomical Observatory of Harvard College* (Cambridge: Wilson)
96. Lassell W *Op. Cit.*
97. Smyth C 1858 Astronomical experiment on the peak of Teneriffe (*sic*) carried out under the sanction of the Lords Commissioners of the Admiralty *Phil. Trans.* **148** 465
98. *Ibid.*
99. Secchi A 1855 Ricerche sopra il planeta Giove, fatte coll'equatoriale de Merz all' osservatorio del Collegio Romano durante l'anno 1850 *Il Nuovo Cimento, Giornale di Fisca, di Chemica, e Delle Loro Applicatione* **2** 351
100. See, for example, Secchi A 1864 Observations of the Spectrum of Jupiter *The London, Edinburgh, and Dublin Philosophical Magazine and Journal of Science* **28** 486
101. Secchi A 1877 *Le Soleil* 2nd edn (Paris: Gauthier-Villars)

Figure 7.1: Joseph Baxendell. (Courtesy of the Royal Astronomical Society.)

refractor at 305× and 223×, respectively. Their drawings represent the marking as a diagonal dark streak extending northward through the NEB. During its observed history, the oval gradually elongated in the **preceding** direction while its limits in latitude remained constant. The streak also became broader and darker. A dark spot at the north end of the Oblique Streak increased in size and darkness until it was resolved into two and then three separate spots[4].

In Baxendell's pictures, the convective element itself can be seen clearly on the northern limit of the NEB[5]. (The Great Red Spot Hollow straddles the central meridian.)

On one night in March, a dark curved marking appeared to connect the Oblique Streak with a dark spot on the edge of the SEB[6]. While an EZ(N) plume may have rested below the Streak, this certainly was an attempt to visually connect unrelated phenomena.

Small projections from the SEB into the equatorial zone during this time also were mentioned[7]. These may have corresponded to the chevron-like markings that originate on the south border of the EZ.

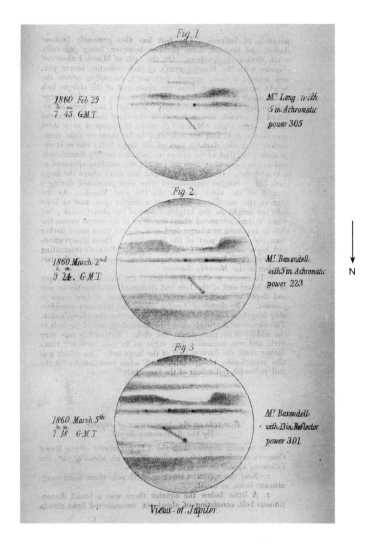

Figure 7.2: Jupiter, 29 February 1860, by Long, and 2 March and 5 March 1860, by Baxendell. (From a plate accompanying 1860 *Monthly Notice of the Royal Astronomical Society* **20**; courtesy of the Royal Astronomical Society.)

A 12 March illustration, made with a $9\frac{1}{2}$ inch aperture refractor at 250×, shows the Oblique Streak to the exclusion of everything else in the northern hemisphere[8]. The Streak is slightly curved, bending toward the East and following the invisible northern boundary of the equatorial zone from mid-disc. The rest of the hemisphere is white. The SEB is stylized to represent partial cloud cover. The Great Red Spot Hollow indents it and sits north of a featureless STrZ.

By 9 April, the Oblique Streak had lengthened to the point that it extended across the planetary disc. (Such streaks may wrap around the entire planet on this time scale.) By this time the preceding end had become especially dark. It was described as 'bluish-black' when contrasted with the 'dull yellowish red colour of the large belt [SEB]'[9].

Long's and Baxendell's observations were substantiated by a report of the Astronomer Royal who described Jupiter as seen on 26 March 1860, through the Royal Observatory's Equatorial: 'In the region above the equator there was a double belt, inclined perhaps 15° to the equator...'[10]. He stated that its outline was irregular and 'had a most decided reddish tinge; a pale brick-red'[11].

By 6 May, the Streak had begun to wane and is hard to see in a drawing by Long, even though seeing conditions allowed him to use 374×[12]. (The Great Red Spot Hollow may lie to the east in the picture; a slightly earlier Baxendell drawing shows an NEB 'barge'.)

In a representation from the next apparition, the Oblique Streak still barely can be detected. It is impossible to say, though, whether it is, instead, a manifestation of another, later NEB feature. This illustration adorns Canadian Simon Newcomb's (1835–1909) *Popular Astronomy*[13].

Diagonal markings observed on Jupiter are of interest because they must interact with the east–west jovian wind field. While the Oblique Streak is the most prominent 19th-century example of such a phenomenon, additional specimens exist[14]. For instance, several beltlike structures appear to be inclined to the equator in 1873 drawings[15].

Northern hemisphere features are more ephemeral on Jupiter, and it is understandable why the recognition of features in the NEB occurred relatively late. Extended markings originating in the NEB are similar to those at the northern limit of the equatorial zone.

What was the Oblique Streak? A column of steam? Berliner Johann Zöllner (1834–1882) pointed out that a steam plume would appear white; a cloud of steam looks grey only in *transmitted* light. Volcanic ash? But one would expect that material from such point sources would spread out and become diffuse as it travelled from the source. The Oblique Streak did not. Maybe the Streak was a break in the clouds, the long sought-after glimpse of the elusive jovian 'surface'[16]?

In actuality, the Streak was likely a product of a convective region at the northern boundary of the NEB. Such convective elements are fairly common. They introduce white material into the planetary wind field south of the element. As this material spreads out, it alters the albedo of the NEB. The **following** edge of the 'streak' is simply the contour of highest contrast between the new material and the NEB. In other words, the dark diagonal is the 'background' cloud feature and not the 'foreground'. This was again the case at the time of Voyager when 'wedge' shapes appeared to be emerging from the NTrZ[17].

Multiple NEB convective sources may exist at any one time. Usually, a second source will appear when an older one is dying, but occasionally they will develop in tandem. Multiple sources enhance the contrast, a fact that may explain the prominence of the particular marking that was the Oblique Streak. Also, the existing high albedo that prevailed over the NEB must have added to the overall effect.

The only unreconcilable fact about the Oblique Streak is its particular angle of slope. The nearly straight depiction of the streak is misleading; a more accurate trace of the maximum contrast contour would probably reveal a more curved marking.

Other, later appearances of convective sources north of the NEB can be found in the drawings of John Birmingham[18], Oswald Lohse[19] and Laurence Parsons[20] from 1870 to 1873. (See chapter 8.)

7.2 The Challenge of Time-Varying Phenomena

In his description of the behaviour of the Oblique Streak, Baxendell noted that its significance lay in its 'showing very strikingly the extent and rapidity of the changes. . . which sometimes take place on the surface, or in the atmosphere, of this magnificent planet'[21]. However, no more was said about the Oblique Streak of 1860 in the literature. We have only the words of William Huggins: 'In addition to the ceaselessly varying appearances of the surface of Jupiter, there would seem to be great periodic changes occurring in the conditions on the planet, of whatever nature those conditions may be, which give rise to the phenomena visible to us'[22].

(Realizations such as this were beginning to interfere with the process of recording jovian events. One observer summarized the difficulties involved in producing his pictures of Jupiter in this way:

> There is, however, very great difficulty in making drawings of Jupiter, in consequence, partly of the overwhelming amount of details to be seen, and partly on account of the short period allowed by the rapid axial rotation of the planet for their delineation. For unless the drawing be completed within twenty to thirty minutes at most, the change of aspect will have become so great, in consequence of the foreshortening and disappearance of some parts, and the fresh appearance and the elongation of others, that no reliable result will be obtained. And it frequently happens, moreover, that observations begun most auspiciously are rendered useless by the definition suddenly becoming bad and continuing so during the short interval at command; and by the time the atmosphere has quieted, the planet may have rotated so much that it is necessary to recommence, to be disappointed possibly again and again[23].)

After the excitement of Lassell's spots and the Oblique Streak, there were

N

Figure 7.3: Jupiter, 21 March 1863, by Gorton. (From Chambers G 1878 *A Handbook of Descriptive and Practical Astronomy* 3rd edn (Oxford: Clarendon) p 113.)

some who believed that the study of Jupiter waned[24]. This lapse of curiosity was, though, to be a brief one.

7.3 Awareness of Colour on Jupiter

Colour became a prominent theme in reports of physical observations of Jupiter during the 1860s. Previously, comments on colour had been sporadic, one of the few examples being Johann Schröter's late-18th century remark that the equatorial zone looked yellow[25]. (All original copies of Schröter's work were lost when his private observatory in Lilienthal was looted and burned during the Napoleonic Wars.) 'Brown' and 'blue' first entered the Jupiter lexicon in the 1830s[26].

I already have mentioned Baxendell's and Airy's asides concerning colour[27]. Warren De La Rue described the jovian colours in more detail. The 'southern belt' and 'northern [belt]' were brown; the poles were yellow—more so in the north[28]. He also mentioned faint yellow streaks appearing under the 'northern belt,' presumably in the northern component of the equatorial zone[29].

During the 1860s, more and more the bands and features of the planet were described routinely in terms of tint and hue. In 1863, John Phillips (1806–1874) presented several drawings to the Royal Society. He praised his achromatic refractor as superior to contemporary instruments in colour definition. In his report to the Society[30], he spoke of 'red' belts, 'grey' polar regions, and 'white and silvery' zones.

For comparison, Phillips said that no such degree of colourization exists on the Moon, but rather that the tint of the belts resembled those of the

Figure 7.4: John Browning. (Courtesy of the Royal Astronomical Society.)

markings on Mars. Phillips, an Oxford professor of geology, speculated on jovian geography by saying 'In fact, it was suggested to my mind that these coloured extra-equatorial belts were land, seen between white clouds, of which the brightest band was on the equator'[31].

Colour was what drew John Browning's (1835–1929) notice to a major reddening of the equatorial zone, which occurred in 1870. Browning's name first comes to our attention in 1868 when he submitted two drawings of Jupiter to the Royal Astronomical Society, made in September and December 1867. These representations were done at the eyepieces of relatively large-aperture instruments: the first, a Barnes (Captain Edward) $10\frac{1}{2}$ inch reflector at 200×, and the second, Browning's own $12\frac{1}{2}$ inch reflector at the same magnification. Browning also described the basic features observable at the time of his drawings: a bright EZ and dark NEB with a 'uniformly corrugated appearance on the lower or northern edge'[32].

Browning knew that most markings on Jupiter remain unchanged on a time-scale of least several months; however, he noted that in the standard De La Rue illustration of 1856, the same corrugated appearance exists[33]. (This aspect of the drawing cannot be seen in the comparatively crude woodblock prints; Browning evidently had access to a version closer to the original engraving.) He inferred that the longevity of this border marking may have physical significance regarding the actual 'surface' of the planet or, at least, is a unique property of that portion of the jovian atmosphere.

Based on Browning's 1868 article, we can conclude three things that made him a likely candidate to report on a major change on Jupiter. To begin with, he was obviously familiar with the existing belt/zone pattern that had been in existence long enough by this time to be considered the norm. Also, he was attuned to the exceptional nature of rapid, semiglobal changes on Jupiter. To these attributes can be added Browning's talent at exposition. From him, for instance, we read of zones 'greatly resembling the colour of electrotype gold'[34]—a vivid, if now anachronistic, description.

7.4   Mayer's Ellipse

The first person to call attention to a change in the jovian *status quo*, however, was Alfred Mayer (1836–1897), a professor of physics and astronomy at Lehigh University[†]. During the autumn of 1869 and winter of 1870, he found the colours of the jovian disc to be unusual and certain belts to be formed oddly. Using a five inch Clark[‡] refractor[35], he spied a 'ruddy elliptical line lying just below the S. equatorial belt'[36]. This feature became more distinct as it approached the central meridian.

The night of 5 January 1870 was one of exceptional seeing, which allowed Mayer to separate Sirius from its companion and other double stars (a test of resolution). On this night he watched the elliptical feature cross the disc until it was bisected by the western limb and then faded. Believing that no one had seen such a marking on Jupiter before, Mayer prepared sketches at 288×. The next day he painted a unique water colour based upon these sketches. This painting was reproduced *in colour*, within the pages of the *Journal of the Franklin Institute*[37] by a chromolithographic process. It represents the first published colour picture of Jupiter.

Mayer figured the drawn shape of the projected jovian ellipse by direct measurement and comparison with calculated values. His measure of the semimajor-axis/semiminor-axis ratio for the elliptical feature (elongated in longitude) was 1:1.51; taking the inclination of the planet into account, this led to a corrected ratio of 1:1.57 for a hypothetical observer directly above the feature. Mayer even noted the phase effect—Jupiter was nearly at **quadrature**, which precluded illumination and, hence, drawing of the extreme eastern limb[38].

Mayer proposed that the southern elliptical feature was a great mass of gas arising in the equatorial region and sweeping southward. He compared it with a terrestrial cyclone in the Earth's southern hemisphere and explained the ellipse's 'flattened' nature by the high rotation rate of the planet[39]. Indeed, the phenomenon seemed to be flattest on the north side, just where the angular

---

† Bethlehem, PA, USA
‡ Alvin Clark (1804–1887) was the first prominent American telescope maker.

Figure 7.5: Alfred Mayer. (From a plate accompanying 1916 *National Academy of Sciences Biographical Memoirs* **8**.)

velocity of a southern object fixed to the surface of a rotating sphere ought to be greatest.

Mayer's ellipse is a pale yellow colour in the chromolithograph today. After being printed, it faded in the ensuing century. The original colour (and likely true colour of the feature) probably is best described contemporaneously by Camille Flammarion (1842–1925) as follows: 'Le 5 janvier, on voyait sous le bord austral de la bande l'équatoriale une longue ellipse *rougeâtre* propuisant l'effet d'une ligne de vapeurs dégagées non loin de l'équateur [my italics]' [On 5 January, one saw under the southern border of the equatorial band a long and reddish ellipse giving the effect of a line of steam emitted not far away from the equator.][40].

This statement is intriguingly similar to those that would be given the Great Red Spot, if Flammarion's ellipse is the same as Mayer's and not a misnomer applied to the general reddening of the equatorial zone (to be discussed later in detail). Both this statement and Mayer's picture put the feature farther north than the current GRS latitude. Neither is of sufficient precision for this discrepancy to be meaningful in itself.

That the elliptical feature was not a product of activity wholly *within* the SEB is clear from the following passage, in which Mayer describes structure in that belt and speculates on both the nature of it and the feature of interest:

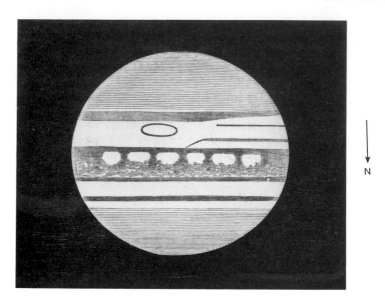

Figure 7.6: Jupiter, 23 January 1870, by Gledhill. (From 1880 *Observatory* **3** p 281; courtesy of the Royal Astronomical Society.)

> The irregular, nearly detached masses forming the S. equatorial dark belt [SEB], with their streamers pointing in the opposite direction of that of the planet's rotation, present appearances which incline one to imagine them distinct masses of cloud or coloured vapor instead of openings in a cloudy stratum disclosing the ruddy body of the planet. The elliptical form below the S. belt also favors a similar opinion...[41].

Here Mayer disagrees with William Herschel and others' interpretation of the darker jovian features as views of the underlying 'surface'.

A picture of Mayer's ellipse is situated next to a late-1870s drawing depicting the Great Red Spot in Proctor and Ranyard's *Old and New Astronomy*[42]. A comparison strongly suggests that Mayer's feature is the great southern feature itself. Its outline is enhanced in this version, however. In the original chromolithograph, it is a most subtle feature.

A similar drawing was made by former schoolmaster Joseph Gledhill (1836–1906). It was drawn on 23 January 1870 and shows Mayer's feature schematically. Gledhill's picture appears in a later paper favouring the interpretation that this is the Great Red Spot[43]. It probably was reconstructed from written notes some time after the fact.

The supposed Great Spot (a near-perfect oval annulus) is in the STrZ. A dark wake follows. At this point the zone expands into the STB slightly[44].

Looking back at pictures made just after or just before Mayer and Gledhill's Great Red Spot candidate was drawn, it is difficult to find a precursor.

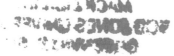

Figure 7.7: Jupiter, 9 October 1869, by Browning. (From 1870 *Nature* **1** p 139.)

Only three suspects, or at most four, can be identified. In Browning's October 1869 drawing, the STrZ slopes somewhat at the western limb and may betray the GRS Hollow hovering there[45]. A year later, the SEB is shown as dark, but tapering to white in the East (the effect of which is to make the STrZ appear broader), suggesting, again, the Hollow[46]. A 14 December 1870 drawing by Thomas Webb has the SEB(S) strangely kinked at mid-disc[47]. This, too, could be caused by the Hollow 'squeezing' into the adjacent band. In the one other possible example, a drawing by Laurence Parsons made on 7 February 1872, a broadening in the SEB(S) merely hints at a Great Spot Hollow[48].

Nowhere else is there mention or depiction of an oval spot. In retrospect, it seems unfortunate that this feature appeared at nearly the same time new activity broke out in the equatorial zone (then considered the more spectacular occurrence), since it divided the attention of Jupiter-watchers. Monitoring of the ellipse was effectively interrupted by these developments in the EZ.

7.5 Initial Reports Focusing on the equatorial zone

Though some astronomers noted an equatorial reddening on Jupiter as early as November 1869[49], Mayer gave one of the most vivid accounts of the 'irregular and violent changes' wracking Jupiter in 1870[50]. On 8 January, he saw that the southern border of the 'southern equatorial band [EZ(S)]' had divided in two[51]. This division increased to a separation of eight degrees of latitude to the south by the 21st. By now the SEB and EZ(S) 'presented the appearance of irregular massed cumulus clouds forming three distinct aggregations...'[52]. At the same time, the EZ(N) appeared canonically to Mayer as consisting of 'nimbus clouds from which drooped streamers which extended to a short distance in the direction opposite to the rotation of the planet'[53].

Turning to the subject of colour once again, to Mayer's eye, the disc of Jupiter was normally a 'light yellow', with belts of 'brownish-grey' crossing it[54]. Sometimes they nearly approached a 'rose colour'; but at other times, the brown in the belts was altogether absent[55]. Then these bands were nearly grey.

In Mayer's painting, the disc is only mildly yellow, whereas between the equatorial belts it is yellow mixed with 'crimson lake'[56]. (As was customary at the time, Mayer expressed the colours of Jupiter in terms of the paints he selected to represent them.) The two visible northern belts and one southern belt are even more strongly crimson lake and approach a 'coppery hue'[57]. The polar regions are yellow with a hint of 'light lead' added[58].

Mayer promised to continue his observations and water colours of Jupiter and to report on them, but the next year he left Lehigh University to start the Department of Physics at the new Stevens Institute of Technology†. Indeed, Mayer was foremost a physicist. He was recognized for his work in acoustics, and for experiments with magnets floating in a magnetic field, which were eventually to have application in atomic physics. Mayer's four-year tenure at Lehigh was seemingly the only one of the several academic positions he held in his career that called for him to undertake astronomical pursuits.

He took this responsibility seriously at the time. Only the year before Mayer published his Jupiter paper (in 1869), he successfully photographed a total eclipse of the Sun for the Nautical Almanac Office, but neither before nor after did he publish additional observations of Jupiter. Thus this promising observer vanished from the planetary scene.

Fortunately, Browning took up where Mayer left off. Already as early as October 1869 he had recorded that the EZ changed from the brightest white portion of the planet to a band of 'strong, greenish yellow'[59]. Browning was the first to attribute significance to the unusual colour of the equatorial zone. The turnabout in colour and albedo inversion caused the EZ to become darker than either the NEB or SEB. On 9 October, 'The color is almost exactly that known to artists as yellow lake', wrote Browning[60]. When observed two days later, the colour was identical; however, there were now white spots within the Zone. On this evening, Richard Proctor (chapter 9) joined Browning at the telescope and confirmed Browning's colours, except red in the belts, which Proctor could not see.

In all of his communications on the subject, Browning seemed intent on getting others to confirm his observations. For instance, he quoted a letter

† Hoboken, NJ, USA.

he received from 'Mr. Brindley of Lewisham' who, with his $8\frac{1}{4}$ inch reflector, observed Jupiter at the same time, independently of Browning: 'I was most agreeably surprised; the belts were so different to what I had seen them before. The dark ones of a dark lake colour, the bright one of a lovely tinted green'[61]. The descriptions are quite similar.

On one night in November (1869), three separate observers drew attention to the equatorial zone: Engineer Thomas Elger[62] (1837–1897) ('peculiarly ruddy'[63]), Edmund Salter ('rich *tawny*'[64]), and Gledhill ('Ruddy tinge'[65]).

By January 1870, said Browning,

> The coloured belt on Jupiter to which I have recently had the honour of calling the attention of the Society [RAS], has undergone many changes both in form and hue since I last described the appearance of the planet. The ochreish-yellow colour is rather fainter, and of a duskier hue, and it is confined to the northern part of the equatorial belt [EZ(N)], instead of covering the whole of it as was formerly the case.... It would, perhaps, be more correct to say that there are two belts near the equator, one to the north [EZ(N)], of a faint dusky yellow, not dark, but very dim, and one to the south [EZ(S)], which is pure white. This belt [EZ(S)], is by far the brightest portion of the planet's disc, and it is the only portion of the disc which is colourless.... The tawny yellow colour now again extends over the whole of the equatorial belt [EZ], which is broader than I have ever seen it before[66].

Browning concluded his March paper 'As this striking outbreak of colour appears to be on the increase, it is very desirable that those observers who have a good western view should observe the planet, and report their observations at every possible opportunity'[67]. This last comment reflects the fact that Jupiter was low in England's sky at the time. Elsewhere, Browning complains that his reflector was restricted to 140× because of the high air mass associated with the planet's low altitude[68].

The 1869 reddening was acknowledged by half a dozen observers of Jupiter familiar to Browning. They employed reflectors from eight to twelve inches in aperture and four to eight inch achromatic refractors. (The reddening was apparently less noticeable through the refractors.) All concurred that the equatorial zone had assumed a yellow, ochreish, brown, or tawny colour, as opposed to its previous pearly white[69].

Not everyone agreed that the 1869 reddening was so dramatic, though. In his report to the Board of Visitors, the Astronomer Royal (Airy) compared a recent drawing by James Carpenter (1840–1899), created at the Royal Observatory, to one made eight or nine years previously. He commented that there appeared to be no significant change in colour[70].

Browning strongly disagreed. However, he proposed that the change in colour may be periodic and that Airy's references might represent the repeat of a particular aspect of a cycle. During the 1868–69 apparition, then, a dissimilar aspect of the cycle presented itself[71].

(In fact, Angelo Secchi, who began observing Jupiter in 1860, had observed a faintly redder-than-usual equatorial zone nine years previously! He found the 1870 reddening to be much more impressive, though[72].)

After responding to the Astronomer Royal[73], Browning learned of Mayer's[74] observations. In a further note, he quoted Mayer on the unusual colours of the equatorial zone and stated that Mayer's and his sets of observations were well correlated.

Browning summarized his observations of the 1869–70 reddening, as reported to the RAS, in an article for *Nature*[75]. There he dwelt on the great variety of colours evident—greater than that previously witnessed. Browning also spoke of an increased number of observable bands. Of the equatorial zone itself, he said, 'Usually it has been free from markings, now it is often covered with markings, which resemble piled-up cumulus clouds...'[76].

Browning included a woodcut of a drawing apparently produced on 9 October 1869 (though the text and figure caption differ as to the exact date). Three southern and two northern belts are shown; the colours are given as 'deep rich yellow [EZ]', 'darker tint of the same colour [NEB and SEB]', 'ashy blue [Polar Regions]', 'pearly white [STrZ and NTrZ]' and 'coppery red [STB and NTB]'[77].

7.6   A Detailed Description of Jupiter and 'Portholes'

Another person observing Jupiter during the early 1870s was the Reverend Thomas Webb (1807–1885). Struck by the transitions he witnessed there, Webb undertook a series of observations from which he compiled a comprehensive picture of the planet's then-current state. However, while Browning emphasized colour, Webb was as much concerned with morphology, of the EZ and all the other belts and zones. Webb was interested in the broader question of what causes (or caused) the distinct qualities of the jovian planets in general. As he put it,

> An important distinction has been repeatedly pointed out, between the group of interior and exterior planets... Either group, as far as observation extends, or fair analogy will carry us, has a character peculiarly its own; the outer being distinguished from the inner by inferior comparative density, but superior magnitude, velocity of rotation, and attendance of satellites. These remarkable differences, though increasing the interest of such researches as may be permitted to us by the Great Ruler of the universe, add materially to their difficulty; and we find it impossible to carry on to remoter planets the analogies which have apparently served us so efficiently in the case of

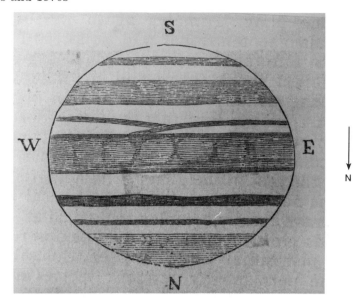

Figure 7.8: Jupiter, 14 December 1870, by Webb. (From 1871 *Nature* **3** p 430.)

our closer neighbour Mars. It is fortunate, therefore, for the purpose of our study, that Jupiter, the nearest at once and the largest of that external group, presents a disc so broad and so luminous as to invite examination even with telescopes of moderate size...[78].

(This philosophy will be expanded upon in chapter 9.)

Webb provides the most extensive word-picture of Jupiter to date. Webb observed Jupiter with his silvered† Newtonian between 15 October 1869 and 11 March 1870 (a period nearly commensurate with Browning's reported observations). This series covered 40 nights at the telescope, 'though sometimes to little purpose' because of obstinate seeing[79]. Webb's telescope was manufactured by With and George, at $9\frac{1}{3}$ inches in aperture, was resolution tested by splitting the double stars τ Andromedae and μ Böotis. Under reasonable conditions, a magnification of 212× could be employed.

Webb noted alterations in the albedo of the belts taking place on the order of two hours. However, neither the physical mixing rate at the available resolution, nor the cooling rate at the temperature of the upper jovian cloud deck, allow albedo changes on this short a time scale. These fast variations must have been due to seeing or simply the rotation of inhomogeneous longitudes onto the visible disc.

† American Henry Draper (1837–1882) was probably the first to observe Jupiter through a silvered-mirror reflecting telescope (1860s). He found the planet 'covered with belts up to the poles' (King H 1955 *History of the Telescope* (Toronto: General Publishing)).

Figure 7.9: Thomas Webb. (Courtesy of the Royal Astronomical Society.)

The varicoloured equatorial zone† which attracted Webb's attention, was 'ruddy or brownish-yellow'[80]. Grey shadings projected from the edge of the SEB, usually perpendicularly to it and sometimes deflecting eastward. Webb estimated that 16–18 such loops girdled the planet, though several were darker or broader than others. To Webb, this gave them the look of an arched bridge wrapped around the EZ, because the southern 'base' of each loop was wider than the top. These 'arches' evolved by November into discrete hollows set slightly obliquely. The 'solid ellipsoids seemed to stand out of... a depressed channel... or it might be compared to a modification of the moulding known as "bead and hollow" in architecture'[81]. These ellipses intruded into the

† Other references to the colour of the EZ, during 1870, include 'The space between the equatorial belts [NEB and SEB] is rose-coloured, becoming of a lighter tint toward the centre'. (Birmingham J 1871 Reports of observations made from 7 January to 6 February 1870, inclusive *Astronomical Register* **8** 84), 'The equatorial belt of a fine copper colour. The other belts grey'. (Whitley H 1871 Report of observations made from 7 January to 6 February 1870, inclusive *Astronomical Register* **8** 85) and '...it [the EZ] appeared to be a dusky orange hue' (Elger T 1871 Report of observations made from 7 February to 6 March 1870, inclusive *Astronomical Register* **8** 94).

SEB up to one-half of that belt's width. (Compare these SEB features with those reported by Browning[82].) This trick of the eye continued. Eventually, Webb described the markings as separated ovals (not unlike Mayer's feature) and saw up to four at once. However, as Jupiter's distance from the Earth increased, the details became harder to see.

(Webb's description is remarkably similar to that of the Astronomer Royal in 1860. Airy saw six clouds in a discontinuous equatorial belt, each roughly circular[83]. He wrote that similar features were visible in Piazzi Smyth's Tenerife pictures, but today they are not obvious.)

To explain these curious equatorial markings, let us return to Browning's original published representation from October 1869. In that drawing, curved intrusions of the dusky SEB interrupt the southern component of the equatorial zone. These 'notches' are caused by large cyclonic elements distributed semiuniformly through the SEB(N). The dark cells appeared connected together and so were not treated as spots, but simply as a belt's irregular edge. Many other prominent observers drew 'notches' after Browning[84].

In the illustrations by Webb for his 1870 article[85], the circular features (of 16 November 1869), in the southern component of the equatorial zone, have come to be called 'portholes'. 'Portholes' represent the same phenomenon as SEB 'notches'. Here, though, the artist's eye has been drawn to the 'curvy' nature of the EZ just above the notches. It takes little misleading of the eye to 'complete' these curves into ellipsoids. These two different representations of equivalent phenomena are precipitated by an ambiguity often purposefully introduced into artwork—that is, the ambiguity betwixt field and foreground. Most Jupiter-watchers preferred to think of temporary features on the planet as being superimposed on a tenebrous background. Hence the 'porthole' was the feature, not the dark cell (individual 'notch').

An Irish country gentleman named John Birmingham (*circa* 1816–1884) wrote a number of papers on solar-system bodies during the 1870s. For one, he constructed a series of 32 illustrations that commenced in the last days of 1869 and continued through early 1871[86]. Most of his drawings of Jupiter show the notched SEB(S) contrasted against a relatively light SEB(N). There are likewise EZ(N) plumes in the drawings, but Birmingham was led to render 'portholes' just once.

'Portholes' show up as well in the artwork of Browning[87] and of Joseph Gledhill[88] and Étienne Trouvelot[89] (chapter 8) between 1870 and 1873.

Among those agreeing that Jupiter was exhibiting extraordinary colours at this time, there were differing opinions as to what these colours really were. Webb addressed this issue in a conciliatory manner, as it had supposedly led to some controversy. He objected to statements made about the 'colour blindness' or 'colour perversion' of certain astronomers[90]. He suggested that the more-important variation in physiological response to what exists on the jovian disc is rather in respect to the *type* of things noticed. That is, certain persons may be more sensitive to seeing differences in albedo or colours, while others are better able to resolve lines and edges. Webb believed that the

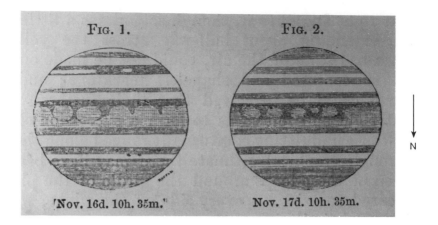

Figure 7.10: Jupiter, 16 November and 17 November 1869, by Webb (From 1870 *Popular Science Review* **9** p 129.)

more-experienced observer may be more attentive to those details that are significant and which differ from the norm. Because of this, a person who has studied Jupiter for a while has an advantage.

Webb made it clear that *he* was one of these individuals. (Webb did not see Mayer's great ellipsoid and apologized by saying that he may have not recognized the anomaly because he was *too* familiar with the planet!.† For instance, he warned that common map projections of the Earth cause people to forget feature foreshortening on the limb of a real globe[91]. As foreshortening changes as the planet rotates, it is necessary to construct the dimensions of a drawn feature properly for a specific instant in time. Features are studied best near the centre of the disc as they also tend to fade near the edges. Webb suggested that this could be due to an envelope of haze about Jupiter but that the effect might be improved upon by better optics. (A haze layer has been discovered high above the visible clouds, though a Voyager occultation experiment was necessary to detect its extent[92].)

Webb complained that probably only a few astronomers really drew well. 'It is much to be regretted', he avowed,

> ...that a certain amount of artistic skill is not considered absolutely necessary in a liberal eduation; the advantage and pleasure derivable from it in after life are so obvious that it may well be questioned whether some of the time that is spent in mastering classical and mathematical niceties of an extremely unserviceable and unpractical nature, might not be better expended in the acquirement of the most useful art of design[93].

† He finally saw it in 1871. (Webb T 1871 Observations on Jupiter in 1870–71 *Popular Science Review* **10** 276.)

Webb cautioned restraint in claiming to portray Jupiter accurately in a depiction. At the same time, he acknowledged that a timely crude sketch may capture a new phenomenon, or one not to be seen again, and has its place alongside more thorough works. Webb considered De La Rue's 1856 rendering the best hitherto published picture of Jupiter[94].

Additionally, Webb discussed instrumental effects in the rendering of colour[95]. Entering into the debate on the advantages of the silvered reflector versus the achromatic refractor, he proposed that for different reasons each instrument produces equal colour distortion. Blue light led astray in the achromatics of the time also was absorbed by the silver of a mirror; therefore, the images produced by both suffered from an overabundance of red. (He did not, though, conclude that this in any way biased observation of increased intensities of orange and red, then lately made on Jupiter.) Webb finally came out on the side of the reflector, which, at least, eliminated the blue altogether, rather than producing an annoying blue fringe. He cautioned, though, that individual achromatics differed in the arrangement of their correcting lenses and that the eyepieces of all instruments introduce effects of their own.

Webb did not mention the use of coloured filters by any observer of Jupiter[96]. This is consequential concerning the pivotal issue of colour distortion but, ironically, suggests that it was not yet recognized that such filters can actually aid the discernment of certain albedo features.

7.7 Continued Reports Focusing on the equatorial zone†

In the 1870–71 apparition, the equatorial zone was more 'ochreish or tawny' than during the previous one[97]. No longer could it be attributed to a terrestrial atmospheric effect or some other fleeting cause[98]. (Besides, no other planet in the sky looked unusual; neither did the Galilean satellites.)

Browning produced two drawings made on 24 and 25 October 1870—two nights of exceptional seeing. By making drawings at the same time on successive dates, nearly all longitudes may be represented. Again, the NTrZ was the brightest zone, and all the northern belts were dark brown (but with less 'copper colour' than previously)[99]. The southern belt/zone structure was

---

† Photography is outside the scope of this work. However, Browning did consider photographic evidence for his EZ 'reddening'. He reported that

> ...Lord Lindsay has shown me two photographic negatives of Jupiter, taken at Mr De La Rue's observatory, within a quarter of an hour of the time I made my first drawing. It is worthy of remark, that the equatorial belt [EZ] in these negatives is almost absolutely transparent: the light from this orange-coloured belt has failed entirely to act on the sensitive collodion surface. I have seen negatives of Jupiter taken during previous years in which the equatorial belt had exerted the most action on the surface, giving the belt as quite opaque. (Browning J 1872 On a photograph of Jupiter *Astronomical Register* **9** 15.)

This evidence vouched for the zone's low albedo, if not its colour.

N

Figure 7.11: Jupiter, 24 October 1870, by Browning. The bottom drawing is from a photograph. (From 1870 *Monthly Notices of the Royal Astronomical Society* **31** p 35; courtesy of the Royal Astronomical Society.)

not as well defined. A sole belt (the SEB) was mottled with 'white cloudy markings or patches'[100]. A series of 'patches' was separated by darker regions. The sequence apparently continued around the planet.

Professor Maria Mitchell (1818–1889), the first female professional astronomer in America, painted a slightly different version of the 1870–71 apparition. Her drawings were checked against those created by three of her Vassar students. Compared to the previous year, 'the rosy tinge of the equatorial belt [EZ] was less marked', according to Mitchell[101]. The SEB during January 1871 had a 'violet tinge', and there was a short, broad 'violet marking' near the northern border of that belt[102]. (Flammarion thought the 1871 SEB had a 'teinte empourprée' [crimson shade][103].)

Figure 7.12: Maria Mitchell. (Courtesy of Pilgrim New Media†.)

Instead of unprecedented red, Mitchell saw numerous white spots at those equatorial latitudes. Some of them were reported to undergo unbelievable gymnastics. On 7 January, a white oval spot was noted preceded by two brown spots. Within an hour, the white spot was situated *between* these brown spots. Later that month, two white spots appeared to change their relative positions in the same length of time!

Mitchell's observations are in the log style of a frequent observer. However, for nearly twenty years prior to 1868, she had been a computer and compiler of ephemerides for the United States Nautical Almanac Office. While both these tasks demanded great accuracy, neither required her to make observations. Mitchell (and many other female astronomers) had infrequent opportunities at a telescope. In this case, a longer time line of observations would help in assessing the character of her descriptions, which seem to include a higher occurrence of shorter wavelength colours than might be expected‡. Mitchell, unfortunately, is far from being unique in that her published observations of Jupiter were made only during a relatively short interval of her astronomical career.

Late in the 1870–71 apparition, Browning concurred 'that the colour of

---

† From the CD-ROM *A Biographical Encyclopedia of Famous American Women*, copyright 1994 by PNM.
‡ Some people have been proven to have retinal sensitivity that extends into the ultraviolet. (Sheehan W 1997 Personal communication.)

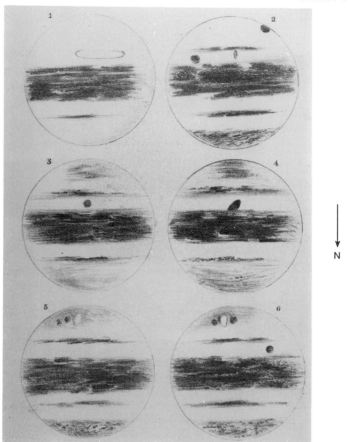

N

Figure 7.13: Jupiter, 1870–1871, by Mitchell. (From a plate accompanying 1871 *The American Journal of Science and the Arts* third series **1**; produced by Aaron Spurr.)

the equatorial belt [EZ] appeared paler than it did last year'[104]. However, Browning discounted the value of observations made when Jupiter was so far from opposition.

The debate as to whether the appearance of the equatorial zone had really changed at all continued into the 1870–71 apparition. At a meeting of the Royal Astronomical Society, everyone present except Francis Penrose (1818–1903) and Arthur Ranyard (chapter 9) maintained that it had not! It was suggested that the new tinctures observed there were the products of the coincidental popular introduction of silvered glass reflectors with large apertures. Browning countered that he had been making observations with his $10\frac{1}{4}$ inch silvered reflector for five years and that until the recent apparition, the 'coppery grey' of the belts and the 'bluish grey' of the poles were the only colours to be discerned—neither of which appeared with the intensity of the 'tawny colour' in the contemporary EZ[105]. Other observers had noticed the coloured EZ with both reflectors and refractors of merely 3–4 inches in aperture.

Indeed, in 1869, Browning had written that colour is seen *best* with small apertures at high magnifications. (He worked with powers of 350–500× under good seeing conditions.) Browning did shift to a $12\frac{1}{2}$ inch aperture telescope in 1867, but, owing to the inferior nature of British skies, usually stopped it down to smaller than 10 inches[106]. Oddly, Browning subsequently contradicted himself and stated that while the colours of stars are displayed sufficiently with small-aperture instruments, he advocated large apertures for seeing the colours of the planets. As usual, Browning cited other observers who agreed[107]. The effect was to leave the question of aperture an 'open' one.

Browning ultimately relied on the sheer vividness of the tones to authenticate them. 'Colour observations are, of course, liable to many sources of error', admitted Browning, 'but though observers may at different times receive different impressions from the same colour, I do not think it possible that any one accustomed to the use of colours would mistake yellow ochre for white'[108]. The NTrZ and STrZ 'always appeared nearly white' and were available as standard comparisons[109].

Browning tried to dispel the myth that Jupiter's rubicund colours result exclusively from conditions in the Earth's own atmosphere. He felt that they were clearest when the atmosphere was particularly *free* of dust or mist, known reddening agents[110].

(Spectroscopists at the Great Melbourne Reflector thought they could see changes in Jupiter's spectrum due to the EZ reddening[111]. One visual observer would afterward claim that in 1870 the hue of the equatorial zone was so intense that it caused the entire disc of Jupiter to appear red to the *naked eye*![112])

The existence of the equatorial zone reddening appears to have ceased to be controversial with the observational support of venerable planetary observer William Lassell, who had been elected president of the RAS in 1870. Said Lassell in *Monthly Notices* 'I acknowledge that I have hitherto been inclined to think that there might be some exaggeration in the *coloured* views I have lately seen of the planet; but the property of the disc in the view I am describing was so unmistakable, that my skepticism is at least beginning to yield'[113].

Lassell had practiced his artistic skills throughout the 1860s as he discovered and recorded various new nebulae. His telescope of choice was still the 24 inch **speculum**, which he defended against what he considered still imperfect achromatic refractors. It was with this instrument that he watched Jupiter.

In 1872 Lassell published a pencil sketch of the jovian disc made at 220 and 430×, with the novel addition of labels beside each band distinguishing that region's perceived colour[114]. He saw a great variety of colours on the planet, including that in the equatorial zone, but just at high magnifications. (As magnification increases, Jupiter fills a greater portion of a telescope's field of view; thus, eventually, the human eye can adjust its contrast range for a field

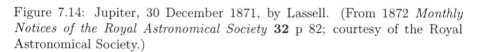

Figure 7.14: Jupiter, 30 December 1871, by Lassell. (From 1872 *Monthly Notices of the Royal Astronomical Society* **32** p 82; courtesy of the Royal Astronomical Society.)

that is predominantly light, as opposed to one that is predominantly black sky.) Regardless, Lassell's credibility by now was based largely on reputation instead of keen observing technique. The elderly astronomer's eyesight had begun to fail. Yet his report was valued highly, and Browning was quick to quote Lassell's testimony.

A vindicated Browning published a colour plate of Jupiter. In describing it, he outdid Mayer in his application of painters' nomenclature: 'In the drawing...the belts were coloured with a mixture of madder-brown and sepia; but two years since, I saw them of a full rich claret colour, the colour of pure purple madder'[115]. From this passage, we see that Browning believed that more than just the equatorial zone had been reddened; the NEB and SEB had, too. The NTrZ was 'rather warmer than ultra marine ash'[116]. As to the colour of the EZ ('Roman ochre' at its ruddiest), Browning conceded that it was fainter now, but blamed this on the less favourable nature of the opposition[117].

Mitchell also described Jupiter as it displayed itself early in 1873. Her straightforward report of observations made at the eyepiece of the formidable 12 inch refractor (200–250×) referred mainly again to spots. As to colour, 'The broad belt [EZ]...was slightly reddish' early in the observations; on 11 March only a 'faint rosy tinge could be seen on the upper part of the broad equatorial belt [EZ(S)]'[118]. Two nights later, 'no rosy tinge could be perceived'[119]. (She did, though, on this occasion notice linear markings in the

Figure 7.15: Jupiter, 26 March, 20 April and 11 May 1873, by Knobel. (From a figure accompanying 1873 *Monthly Notices of the Royal Astronomical Society* **33**; courtesy of the Royal Astronomical Society.)

equatorial zone—a phenomenon mentioned with increasing frequency in the literature.)

Edward Knobel (*circa* 1850–1930), primarily a student of Mars, was unique in compiling a detailed account of the 1873 apparition of Jupiter. Noting this, he presented the Royal Astronomical Society (of which he would eventually be president) with his sketches rendered during that period at an $8\frac{1}{2}$ inch reflector[120]. In the equatorial zone he found 'long, irregular, broken masses, horizontal and inclined at a considerable angle to the equator...'[121]. The 'north temperate dark belt [NTB]' was double now (probably due to the appearance of one of the equatorial belt components); the 'south tropical dark belt [SEB]' was irregular, widening toward the west. As for the EZ reddening, 'atmospheric influences this year have been fatal to observations of colour'[122]. However, he did say that the 'brick-red hint' of the SEB was 'more decidedly red than the darker parts of the equatorial zone [EZ]'[123]. The equatorial zone had clearly lost the hue that had caused such excitement, if not all hints of colour.

(Flammarion considered the zone to be merely 'jaune-cuir [crimson shade]' in 1873[124]. Yet Pietro Tacchini (1838–1901), a colleague of Secchi, was still willing to call parts of the EZ 'de couleur rose'[125].)

After four years, Jupiter had apparently returned to 'normal'[126]. Fittingly, Browning pronounced the end of the colourful equatorial zone phenomenon. By opposition in 1873, most indications of the reddening had subsided. In fact, the EZ was devoid of any features at all. Even the 'copper-coloured' belts were fainter than usual to Browning's eye[127].

As the atmosphere of Jupiter returned to normal, that of the Earth above England turned against Browning and other astronomers: 'During the whole of the opposition the definition has been so uniformly bad that I have found it useless to make drawings of the planet'[128].

7.8   A Disturbance on Jupiter

As has been shown, the albedo of the jovian tropics was of paramount concern during this period. For a time, planetary astronomers—except Mitchell!—seemed to lose their preoccupation with features.

Yet Birmingham had first brought attention to himself by discovering nova 1866 Corona Borealis, while strolling home from a friend's house one night. (It became the brightest seen since 1604.) Birmingham was a man who had a keen sense for subtle changes in appearance; he is known best for a detailed catalogue of red stars. Perhaps it is for this reason that only Birmingham documented, in his collection of drawings[129], a disturbance apparently taking place in the SEB during the 1869–70 and 1870–71 apparitions.

On 18 December 1869, the notches in the SEB observed two nights previously could not be seen. Instead, a thin line appeared. By the 19th, this could be seen to curve. The drawing of 20 December reveals that the convex portion of the line intercepted the STrZ. The mark was likely the head of an SEB disturbance taking place on the planet. Birmingham watched it evolve through 23 February, a busy night on which he made three different drawings of Jupiter[130].

(In the fifteen years prior to the Voyager encounters, three south equatorial belt disturbances occurred on Jupiter; they are not especially rare events. The disturbances appear as they did to Birmingham and others in the 19th century: at a—seemingly arbitrary—longitude in a uniformly whitened SEB, the belt begins to lower its albedo and exhibit structure. This darkening progresses rapidly. Ultimately the entire belt is restored.)

7.9   The Significance of the EZ Reddening

While there exists a body of data from eight or nine observers substantiating the appearance of unparalleled red colouring in the equatorial zone during the early 1870s, in the final analysis of its intensity, instrumental effects cannot be ruled out completely. Browning, upon whom so much of the evidence rests, was foremost a telescope maker. (He may have been sensitive to the fact that he did not have all the credentials possessed by other planetary astronomers of the time; he was quick to fend off criticism of his observations by invoking the supporting opinions of men with greater reputations.) A popularizer of the reflector, Browning was, in fact, the first to produce such telescopes for the mass market: mirrors ground by With mounted in a sturdy-but-affordable mount and base[131]. Yet, for this same reason, it seems unlikely that he would want to foster the impression that his or any reflector introduced spurious colour.

The degree to which Browning *wanted* to see colours on Jupiter is another matter. It should be noted that he initially manufactured spectroscopes[132] and that this vocation suggests an inherent interest in the subject.

Figure 7.16: A Browning Newtonian reflector. (From Sharpless I and Philips G 1882 *Astronomy for Schools and General Readers* 3rd edn (Philadelphia, PA: Lippincott) p 275.)

The inordinate attention given to the low equatorial zone albedo (and its colour) in Browning's time can be explained by the fact that such a thing had not been documented accurately heretofore. However, it has happened many times subsequently. During the 1964 apparition, the EZ was quite dark and orange. The attributed mechanism for this, as well as events nearly 100 years earlier, is the exposure of a reddish chromophore in the jovian atmosphere.

The nature of this lower-level belt chromophore remains unknown. Ammonium sulphides, phosphorus compounds and complex organic molecules have all been suggested as candidates. With each of these, tiny variations in the distribution of a single chromophore could affect the hues and brightnesses seen. Thus, extremely subtle changes can account for the variety of characterizations of observations made, even nearly simultaneously, vigorously expressed and defended during this episode.

For, at the end of the debate, colour remains a perceived quantity, subject always to interpretation by the brain. It has been shown that under strict, photometrically controlled conditions, *nothing* on Jupiter should appear red[133]. The planetary disc becomes, then, really quite bland. Belts are, at most, yellow. Only 'barges', which may require a separate chromophore, are orange or brown. Even the Great Red Spot is red only in the sense that it is highly unreflective at short wavelengths; that is, it is 'not blue'[134].

Planets are extremely bright in the sky; but as image scale increases, their discs expand to fill the detector area (e.g., the eye). At high magnifications, contrast reduces as the object being observed dims. Colour contrast may be increased, though, with wavelength perception being shifted nonlinearly away from the centre of the visible spectrum: dim orange–yellow turns darker orange (i.e., brown), and faint yellow becomes olive green[135]. This processing is intrinsic to the observer. Extrinsically, on top of this, the blue scattering of the Earth's atmosphere introduces a small systematic shift. It is impossible to avoid, then, the fact that colour determination at the eyepiece of a telescope is biased toward the garish.

This is as true today as it was in the last century. Ironically, even our electronic 'eyes' are biased. Voyager spacecraft image processing produced a small characteristic 'red-shift'[136]. Publicity photographic prints released after the Voyager encounters were colour enhanced; in some cases these same prints ended in the hands of scientists studying the planet's albedo!

Why were vivid colours introduced? The answer lies in that they had been so 'observed' for eleven decades. The eye expected them. In an unusual way, colour descriptions, such as those repeated in this chapter, continue to affect the way we choose to 'see' Jupiter.

The change in the EZ of *circa* 1870 was real. How dramatic or singular it was remains uncertain. It is still significant, though, for the reasons cited above, and because it induced observers to discuss for the first time what the colour of the equatorial zone (and other regions on Jupiter) 'should' be†. Moreover, the event excited people and may have indirectly caused more telescopes to be obtained by an increasing number of amateur natural scientists and directed toward Jupiter. (More papers were published on the appearance of Jupiter from 1870–1873 than were published during the previous three decades.)

Also, it should not escape our attention that, before the era of the equatorial zone reddening, colour never was mentioned in descriptions of a large southern oval. After chromatic terms had entered the vocabularies of planetary astronomers, though, when a similar feature was recorded in 1878, it was identified immediately by its colour: 'red'. Would the history of the recognition of the Great Red Spot have been different if this dialogue on jovian colour, and the establishment of expectation of future colour changes there, had not taken place?

---

† Most published observation reports from the mid-1870s onward include some description of colour. In 1876, a group of astronomers soliciting observations of Jupiter stated that 'Careful notes of the tints and colours of the belts are *most* important'. (Quoted by Hirst G 1877 Some notes on Jupiter during the opposition of 1876 *Journal and Proceedings of the Royal Society of New South Wales* **10** 83; my italics.)

7.10 Summary

By the seventh decade of the 19th century, jovian observers had begun to spring from certain common backgrounds and pursued astronomy in a similar way. They were documenting phenomena increasingly more difficult to 'catch', such as low-albedo features that changed markedly with time.

A theme of this chapter has been the perception of shape and colour on Jupiter. The illustrative examples were a large northern convective source and an apparent reddening of the equatorial zone, both witnessed and commented upon by a lengthy list of observers. (The Oblique Streak was the first jovian feature to have a recorded morphology more complex than that of a simple 'band' or 'spot'.) Along the way, a report on a provocative elliptical feature was examined as was an account of another band-extensive disturbance on the jovian disc. These events established the beginning of a new age of thorough amateur and semiprofessional observations of Jupiter made with good quality silver-mirrored telescopes†.

Endnotes

[All titles are written in full, with the exception of '*Mon. Not.*' for the *Monthly Notices of the Royal Astronomical Society*.]

1. Huggins W 1862 On the periodical changes in the belts and surface of Jupiter *Mon. Not.* **22** 294
2. Isaac Long 1861 *Mon. Not.* **21** 101
3. Long J 1860 On the appearance of Jupiter *Mon. Not.* **20** 243; Baxendell J 1860 *Mon. Not.* **20** 243
4. Baxendell J 1860 Observations of the oblique belt on Jupiter *Memoirs of the Literary and Philosophical Society of Manchester* **1** 253
5. Better reproductions of Long's and Baxendell's original artwork appear in Zöllner F 1871 Über das Rotationsgesetz der Sonne und der grossen Planeten *Berichte über die Verhandlungen der Königlich Sächsischen Gesellschaft der Wissenschaften zu Leipzig* **23** 49
6. Baxendell J 1860 *Mon. Not.* **20** 243
7. Baxendell J 1860 Observations of the oblique belt on Jupiter *Memoirs of the Literary and Philosophical Society of Manchester* **1** 253
8. *Ibid.*
9. Baxendell J 1860 *Mon. Not.* **20** 243
10. Airy G 1860 Remarks on the appearance of Jupiter *Mon. Not.* **20** 244
11. *Ibid.*

† It was during this time that cylindrical-projection ('planispheric') charts of Jupiter were first constructed, based on sketches of the jovian disc (e.g., Green N 1873 Planisphere of Jupiter, April, 1872 *Astronomical Register* **10** 169). This technique for portraying the planet continues to be used today; computers now construct such charts based on Voyager, Hubble Space Telescope and Galileo imagery.

12. Baxendell J 1860 Observations of the oblique belt on Jupiter *Memoirs of the Literary and Philosophical Society of Manchester* **1** 253
13. Newcomb S 1889 *Popular Astronomy* 6th edn (New York: Harper). Newcomb was a professor at the United States Naval Observatory.
14. See the extensive discussion in Rogers J 1995 *The Giant Planet Jupiter* (Cambridge: Cambridge University Press)
15. Browning J 1873 Note on the disappearance of the coloured equatorial belt of Jupiter *Mon. Not.* **33** 475
16. Zöllner F 1871 Über das Rotationsgesetz der Sonne und der Grossen Planeten *Berichte über die Verhandlungen der königlich Sächsichen Gesellschaft der Wissenschaften zu Leipzig* **23** 49
17. Terrile R and Beebe R 1979 Summary of historical data: interpretation of the Pioneer and Voyager cloud configurations in a time-dependent framework *Science* **204** 948
18. Birmingham J 1871 Beobachtungen des Jupiter *Astronomische Nachrichten* **77** 301
19. Proctor R and Ranyard A 1892 *Old and New Astronomy* (London: Longmans, Green)
20. Parsons L 1874 Notes to accompany chromolithographs from drawings of the planet Jupiter, made with the six-foot reflector at Parsonstown, in the years 1872 and 1873 *Mon. Not.* **34** 235
21. Baxendell J *Op. Cit.*
22. Huggins W 1862 On the periodical changes in the belts and surface of Jupiter *Mon. Not.* **22** 294
23. De La Rue W 1857 Observations of Jupiter, during October 1856 *Mon. Not.* **17** 5
24. Smyth W 1860 *The Cycle of Celestial Objects Continued at the Hartwell Observatory to 1859* (London: Nichols)
25. Dick T 1838 *Celestial Scenery; or the Wonders of the Planetary Solar System Displayed* (Saint Louis: Edwards and Bushnell) Reprinted in 1854.
26. 1877 Coloured belts on Jupiter *Nature* **15** 282
27. Baxendell J 1860 *Mon. Not.* **20** 243; Airy G *Op. Cit.*
28. De La Rue W *Op. Cit.*
29. *Ibid.*
30. Phillips J 1863 On the belts of Jupiter *Proceedings of the Royal Society of London* **12** 575. Phillips was one of the founders of the British Association for the Advancement of Science.
31. *Ibid.*
32. Browning J 1868 On a persistent marking on Jupiter *Mon. Not.* **28** 213
33. De La Rue W *Op. Cit.*
34. Browning J 1870 Note on the alteration in the colour of the belts of Jupiter *Mon. Not.* **30** 220
35. Gordon R 1997 Personal communication
36. Mayer A 1870 Observations of the planet Jupiter *Journal of the Franklin Institute* **59** 136

37. *Ibid.*
38. *Ibid.*
39. *Ibid.*
40. Flammarion C 1877 *Terres du Ciel* (Paris: Libraire Académique Didier). Flammarion established the French Astronomical Society.
41. Mayer A *Op. Cit.*
42. Proctor R and Ranyard A *Op. Cit.*
43. Gledhill J 1880 Jupiter in 1869 and 1879.—the 'Ellipse' and the 'Red Spot' *Observatory* **3** 279
44. *Ibid.*
45. Browning J 1870 Changes in Jupiter *Nature* **1** 138
46. Browning J 1871 Note on the change in the colour of the equatorial belt on Jupiter *Mon. Not.* **31** 75
47. Webb T 1871 The planet Jupiter *Nature* **3** 430
48. Parsons L *Op. Cit.* A slightly better reproduction of this drawing appears in Proctor R and Ranyard A *Op. Cit.*
49. Flammarion C *Op. Cit.*
50. Mayer A *Op. Cit.*
51. *Ibid.*
52. *Ibid.*
53. *Ibid.*
54. *Ibid.*
55. *Ibid.*
56. *Ibid.*
57. *Ibid.*
58. *Ibid.*
59. Browning J 1870 On a change in the equatorial belt on Jupiter *Mon. Not.* **30** 39
60. *Ibid.*
61. *Ibid.*
62. Fielder G 1963 Some lunar studies by T G Elder *Sky and Telescope* **25** 248
63. Denning W 1871 Report of observations made from September 7 to November 6, 1869, inclusive *Astronomical Register* **8** 15
64. Denning W 1871 Report of observations made from November 7, 1869 to January 6, 1870, inclusive *Astronomical Register* **8** 59
65. Gledhill J 1871 Physical observations of Jupiter, from Nov. 4 to Dec. 31, 1869 *Astronomical Register* **8** 81
66. Browning J 1870 Note on further changes in the coloured belt of Jupiter *Mon. Not.* **30** 153
67. *Ibid.*
68. *Ibid.*
69. Browning J 1870 Note on the alteration in the colour of the belts of Jupiter *Mon. Not.* **30** 202
70. *Ibid.*

71. *Ibid.*
72. Secchi A 1877 *Le Soleil* 2nd edn (Paris: Gauthier-Villars)
73. Browning J 1870 Further note on the change of the colour in Jupiter *Mon. Not.* **30** 220
74. Mayer A *Op. Cit.*
75. Browning J 1870 Changes in Jupiter *Nature* **1** 138
76. *Ibid.*
77. *Ibid.*
78. Webb T 1870 The planet Jupiter, 1869–1870 *Popular Science Review* **9** 127
79. *Ibid.*
80. *Ibid.*
81. *Ibid.*
82. Browning J *Op. Cit.*
83. Airy G *Op. Cit.*
84. See, e.g., Mayer A *Op. Cit.*; Lassell W 1872 Remarks on the planet Jupiter *Mon. Not.* **32** 82; Parsons L *Op. Cit.*; Winlock J 1876 *Astronomical Engravings of the Moon, Planets, etc.; Prepared at the Astronomical Observatory of Harvard College* (Cambridge: Wilson)
85. Webb T *Op. Cit.*
86. Birmingham J 1871 Beobachtungen des Jupiter *Astronomische Nachrichten* **77** 301
87. Browning J 1870 On a change in the colour of the equatorial belt of Jupiter *Mon. Not.* **30** 39
88. Gledhill J *Op. Cit.*
89. Winlock J *Op. Cit.*
90. Webb T 1871 The planet Jupiter *Nature* **3** 430
91. *Ibid.*
92. Cook A, Duxbury T and Hunt G 1979 A lower limit on the top of Jupiter's haze layer *Nature* **280** 780
93. Webb T *Op. Cit.*
94. *Ibid.*
95. *Ibid.*
96. *Ibid.*; Webb T 1871 Observations on Jupiter in 1870–71 *Popular Science Review* **10** 276
97. Browning J 1871 On a photograph of Jupiter *Mon. Not.* **21** 33
98. Flammarion C 1872 Variabilité de l'eclat de Jupiter *Les Mondes* **28** 555
99. Browning J *Op. Cit.*
100. *Ibid.*
101. Mitchell M 1871 On Jupiter and its satellites *American Journal of Science and the Arts* **1** 393
102. *Ibid.*
103. Flammarion C *Op. Cit.*
104. Browning J 1871 Note on the change in the colour of the equatorial belt of Jupiter *Mon. Not.* **31** 201

105. *Ibid.*
106. *Ibid.*
107. Browning J 1872 The condition of Jupiter *The Student and Intellectual Observer* **1** 1
108. Browning J 1871 On the change in the colour of the equatorial belt of Jupiter *Mon. Not.* **31** 201
109. *Ibid.*
110. Browning J 1872 The condition of Jupiter *The Student and Intellectual Observer* **1** 1
111. Le Sueur A 1874 On η Argûs and Jupiter's Spectrum *Transactions and Proceedings of the Royal Society of Victoria* **10** 23
112. Parsons L *Op. Cit.*
113. Lassell W *Op. Cit.*
114. *Ibid.*
115. Browning J *Op. Cit.*
116. *Ibid.*
117. Browning J 1872 On some observations of Jupiter in 1871–72 *Mon. Not.* **32** 321
118. Mitchell M 1873 Observations on Jupiter and its satellites *American Journal of Science and the Arts* **1** 38
119. *Ibid.*
120. Knobel E 1873 Note on Jupiter, 1873 *Mon. Not.* **33** 474
121. *Ibid.*
122. *Ibid.*
123. *Ibid.*
124. Flammarion C *Op. Cit.*
125. Tacchini P 1873 Sur quelques phénomènes particuliers offerts par la planète Jupiter pendant le mois de Janvier 1873 *Comptes Rendus* **76** 423
126. Proctor R 1873 News from Jupiter *Popular Science Review* **12** 348
127. Browning J 1873 Note on the disappearance of the coloured equatorial belt of Jupiter *Mon. Not.* **33** 475
128. *Ibid.*
129. Birmingham J *Op. Cit.*
130. *Ibid.*
131. King H 1955 *History of the Telescope* (Cambridge: Sky)
132. *Ibid.*
133. Young A 1985 What color is the solar system? *Sky and Telescope* **69** 399
134. Beebe R and Hockey T 1986 A comparison of red spots in the atmosphere of Jupiter *Icarus* **67** 96
135. Young A *Op. Cit.*
136. *Ibid.*

Chapter 8

The Modern Discovery of the Great Red Spot

Probably never since Jupiter became an object of telescopic study has more attention been bestowed upon him, or a deeper interest felt in the wonderful changes which are constantly being produced on his surface, than has been created by the advent in 1878 of a tremendous 'red spot' in the southern hemisphere of the planet. This great marking seems permanent, but how long it will last no one can tell. It would not be astonishing news if some fine night it should be missing.

Professor Edward E Barnard
1880

The Great Red Spot is an icon. From computer programs to children's toys, a stylized red spot on an outline disc is enough to symbolize Jupiter and to distinguish it from similar representations used for other planets. Even though the belts and zones are more evident through a telescope than the GRS, the public knows Jupiter as the planet with 'the spot'.

8.1 Observations Made at Parsonstown

William Parsons (1800–1867), Third Earl of Rosse, set out to build the world's greatest telescope. He succeeded. In 1845, Lord Rosse completed a 72 inch aperture reflector, aptly named the 'Leviathan'.

During the heyday of the Leviathan, George Airy mentioned Jupiter in his report on the great instrument: 'The account given by another astronomer of the appearance of *Jupiter*, was, that it resembled a coach-lamp in the telescope; and this well expresses the blaze of light which is seen in this instrument'[1]. The Leviathan was obviously not a planetary telescope; its great aperture was of no significant use and, as implied in this passage, something of a hindrance in observing bright objects. Indeed, the 72 inch's major accomplishment had to do not with the solar system, but with distant and faint 'nebulae'.

Yet late in its history, and after it had been provided with a clock-drive, William Parsons' son Laurence (1840–1908) directed the Leviathan again toward Jupiter. He was induced to do this by the growing interest in Jupiter generated by Browning and others. The Fourth Earl of Rosse believed that the great mirror would reveal detail hitherto unseen. Perhaps, too, he thought it might contribute to the question of colour. (His principal goal was to produce watercolours of the planet). However, two things made this a dubious endeavour: the poor figure of the speculum (compared to a **Foucault-tested** silvered mirror) and local seeing conditions (which limited the resolution of much smaller instruments). The Leviathan did yield detail; an onslaught of fuzzy 'streaks', 'patches', 'markings', 'clouds', and 'spots' overwhelms the reader of the notes accompanying the paintings[2] and no doubt taxed the artists. Still, there is little new information in this mass of overmagnified data (collected at 414× and even 650×, occasionally).

A survey of recognizable features shows that plumes, notches and 'portholes [*sic*]' continued to be depicted[3]. There is evidence for the Great Red Spot Hollow in the drawings from December 1872 and January, February, March and April 1873. (Parsons even used the word 'reddish'[4].) Lord Rosse's artists drew an NEB source, and its resulting diagonal, only once.

Parsons' observations are useful in that he also reported thoroughly his observing procedure and how he produced his artwork. Most earlier published papers expended little print on procedure. Listing the date, time, instrument (usually specified by focal length, not aperture), magnification used and, perhaps, making a brief comment on the quality of the night (usually favourable—there were better things to do on a bad night) sufficed. Even location of the observing site often has to be deduced as the published address of the author. It almost seemed as if there were a race on to reach the descriptions of the observations as quickly as possible, thus sparing us information useful in interpreting the ensuing work. Most regrettable is the fact that rarely was the method of drawing an image of Jupiter at the telescope explained. It was assumed that this was obvious, known or both.

In contrast, Parsons recognized that procedure could affect result. He tells us plainly how he produced his drawings and what the value of taking certain measures was. First, sketches were made quickly at the eyepiece so as not to be distorted by motion. These sketches were made on standardized ellipses calculated from ephemeris data in the *Nautical Almanac*. (Phase effects were ignored). The sketches were then transformed into watercolours, usually immediately the next morning, by 'Dr Copeland'[5], who was guided partly by a list of colour descriptions attached to the sketches. Parsons published these descriptions with the paintings for independent comparison and determination of fidelity. Pigments were mixed freshly each day so as not to

Figure 8.1: Laurence Parsons. (Courtesy of Lick Observatory.)

introduce a colour bias in the painting process. No two renderings were compared until finished. Parsons designed all these steps to insure objectivity and accuracy, and he assured his readers that the published lithographic versions were authentic reproductions of the originals.

Some general trends and phenomena can be made out from Parsons' work. He, too, saw the 'washing out' of the equatorial zone during the 1873 apparition. He also noted markings there, which he used to calculate a rotation period (9 h 55 min 3.8 s). Using a rotation value earlier computed by Johann Schmidt, which Parsons considered most reliable, he then calculated the individual proper motions of three EZ(N) spots[6].

Parsons claimed that a major upheaval took place in 1873[7], but it is unclear whether he meant to associate the whitening of the equatorial zone with this event. This activity shows up in the drawings as a probable south equatorial belt disturbance.

A disturbance actually was documented first by European emigrant Étienne

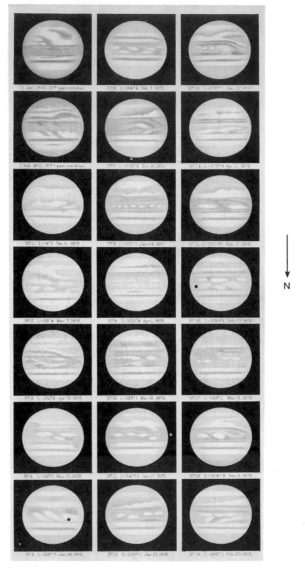

Figure 8.2: Jupiter, 1873, by Parsons. (From lithographs accompanying 1874 *Monthly Notices of the Royal Astronomical Society* **34**; courtesy of the Royal Astronomical Society.)

Trouvelot (1827–1895) at the eyepiece of the 15 inch refractor of the Harvard College Observatory. In a 6 March 1873 drawing, the SEB seems to have a surplus of components[8]. What is happening is more apparent in Parsons' drawing made five days later. By then, the SEB was extraordinary. In the picture, it looks like two oblong loops of dark material linked at mid-disc. The SEB was drawn oddly again on 22 March and 10 April. After this, the belt stabilized for a time into a dark southern and light northern component[9].

While observing at the Leviathan, Parsons documented an astronomical seeing-induced effect very important in the observation of the planet. During nights of otherwise bad seeing, he noted brief instances when the disc of Jupiter would become sharp and well resolved[10]. These momentary conditions of good seeing have since become the targets of photographic patrol observations made of the planets. Multiple short exposures are taken in the hope that at least one image of the set will coincide with one of these unpredictable and rare periods of unusual atmospheric stability. Therefore it will exhibit higher resolution than ordinarily would be expected for the night.

Parsons ceased astronomical observations in 1878. Other astronomers continued to use the world's largest aperture telescope (with increasing infrequency) until it was dismantled in 1913 or 1914[11].

The accuracy of the artwork commissioned by Parsons compares favourably with the written reports of other astronomers observing Jupiter at the same time. But the Earl made one *faux pas* in his 1874 paper. In it he suggested that a drawing made at Parsonstown and another made by François Terby (1846–1911)[12] on the same night, 7 March 1873, differed regarding the appearance of a certain spot. This was because of a one-hour error in Terby's ephemeris, said Parsons. Terby quickly replied by letter[13] from Belgium that this was definitely not so, and Parsons dutifully published the correspondence in a subsequent *Monthly Notices*[14]. He diplomatically proposed that Terby's spot must have been a different one than that depicted in the Parsonstown lithograph and that it must have been very transient in nature. (Terby observed with nothing more than a $3\frac{1}{4}$ inch aperture telescope!) Parsons further pointed out, though, that the feature did not appear in artist Nathaniel Green's (1823–1899) drawing also made on 7 March, either. This minor dispute would not have been possible only a few years before, when multiple drawings of Jupiter made on the same night were exceedingly scarce.

8.2 Observations Made at Moscow

Overlapping Parsons to some extent, another aristocrat, Professor Fedor Bredikhin (1831–1904), was undertaking a parallel project. In seeming isolation from the rest of the European astronomical community, Bredikhin studied comets. He also produced a systematized sequence of jovian drawings (and commentary on them) from the time he assumed the directorship of the Moscow Observatory in 1873. Like Parsons', Bredikhin's concept surpassed its practical substance in importance.

Bredikhin's work was confused not by grossly exceeding the resolution limit—he used a modest 9 inch Merz refractor—but by the generally abysmal observing conditions that he himself acknowledged. (Bredikhin drew with the same eyepiece on which he had mounted a micrometer; his measurements of

Figure 8.3: Fedor Bredikhin. (From the frontispiece, Pskovskii I (ed) 1986 *Istoriia Astronomicheskoi Observatorii Moskovsko Universiteta i GAISh* (Moscow: Moskovskogo Universiteta), courtesy of Robert McCutcheon (Computer Sciences Corporation).)

band widths were no doubt reasonably accurate, but he was thereby stuck with a single 250× magnification that could not be adjusted for seeing conditions.) Also, Bredikhin's colour descriptions are suspect because of the use of a lens objective, though he devoted much print to them. His 'verdâtre' [greenish] NEB and 'violette, avec une nuance roseé' [purplish with a pinkish shade] equatorial zone do little to persuade otherwise[15].

Still, Bredikhin had the right idea. His drawing-by-drawing captions refer to a standard disc model that divides the planet into six (asymmetric) regions, a–f. His micrometer measurements accompany them immediately. Each apparition report begins with a statement about his observatory and instrumentation. It continues with the orientation of the jovian equator with respect to the celestial meridian, an explanation of the nomenclature system in use and a summary of notable events[16].

Bredikhin's morphological descriptions were imaginative and elaborate. Consider this from 15 April 1874:

> La bande obscure avec la série de perles au dessous d'elle peut être comparée à une étoffe de couleur sombre, doublée d'une autre étoffe blanche et luisante. Pour que la comparaison soit plus exacte, il faut adjouter que le bord de l'étoffe est soulevé inegalement dans toute son étendue, et forme ainsi des plis irréguliers, en laissant voir sa doublure luisante. [The obscure band with the string of pearls above it can be compared to a dark fabric, lined with another white and shiny fabric. For a more accurate comparison, we need to add that the border of the fabric is often inequally lifted up over the whole surface, and then forms irregular creases, letting us see its shiny lining][17].

Bredikhin used the expression 'grains des perles' [grains of pearls] to refer to small bright spots[18]. Numerous white spots appear in Bredikhin's reports. It is clear in at least one case that he is talking about Lassell's spots.

Diagonals in the 'zône glaciale' [icy zone] (NEB) appear in Bredikhin's descriptions as 'cornes' [horns][19]. This is a term that is also used to refer to the leading edge of a disturbance during the 1874 apparition. An EZ plume is a 'corne' if seen as the region separating the high albedo feature; it is a 'protubérance claire et arrondie' [a clear and rounded protuberance] if seen as the 'head' of the feature itself[20].

In April, 1875, Bredikhin recorded a large southern hemisphere oval. Inspection of the drawing for the 25th of that month shows that the oval is *prolate*, a shape of low confidence value[21]. Just as curious is a 'langue aiguë' [sharp tongue], emerging northward from the STB in Bredikhin observations of a year later[22].

These unverifiable features make the relatively bland report of the 1877 apparition[23] all the more disappointing. (Considering how obvious the Great Red Spot was to be in less than a year, why did Bredikhin fail to observe it?) While Bredikhin was a more impassioned observer than Parsons, neither man obtained a familiarity with Jupiter necessary to make sense out of the onslaught of potential phenomena waiting there.

8.3 The 1874 Apparition

There were still those with whom Jupiter was a favourite. In April 1874, artist John Brett† (1831–1902) saw some 'white patches' in the equatorial zone. Brett would easily have admitted that this was no longer an unusual observation except that these particular features, to Brett, appeared to cast *shadows*. Or, as Brett put it, '... the light patches are bounded on the side

---

† Brett was known for 'his pictures of sea and coast scenery ...' (1902 'Obituary—John Brett *Observatory* **25** 101)

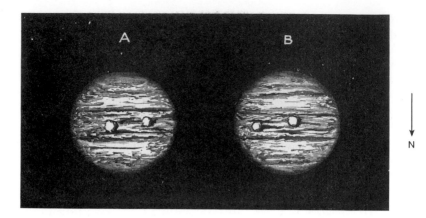

Figure 8.4: Jupiter, 23 and 28 May 1876, by Brett. (From 1876 *Monthly Notices of the Royal Astronomical Society* **36** 356; courtesy of the Royal Astronomical Society.)

farthest from the Sun by a dark border shaded off softly toward the light, and showing in a distinct manner that the patches are projected or relieved from the body of the planet'[24].

Brett found this observation to be a key one with respect to an important question about the solid 'surface' of Jupiter: was it represented by the belts or the zones? (William Herschel had voted for the belts, Johann Schröter for the zones.) Brett felt that these shadow-casting spots suggested that it was *neither*, and that 'all we see of Jupiter consists of semi-transparent materials'[25]. This was because the shadows did not appear to be sufficiently well defined to be falling on a solid 'surface.' The shadows also were not uniform, as expected on an opaque surface, but instead they trailed off toward their edges.

Brett observed his 'patches' through a $9\frac{1}{4}$ inch silvered reflector, usually at $400\times$[26]. Lassell confirmed them with his famous 20 foot focal-length telescope, before opposition eradicated the supposed shadows.

In March 1875, French astronomy-popularizer Camille Flammarion would see spots with 'shadows', too. He carefully worked out the positions of the satellite shadows, none of which coincided with any of these spots. Moreover, the motion of the black areas mimicked those of leading, intrinsic, white spots. He could not reconcile their displacement, though, with the incidence angle of solar illumination. He therefore was forced to question whether they were true shadows. His alternative hypothesis was that they were 'analogue à la pluie suivant un nuage ...' [similar to the rain after a cloud][27]. Both ideas were incorrect.

With few exceptions, continental astronomers of this time remained measurement oriented—they had not adopted the British habit of descriptive morphological surveillance. Flammarion found and drew many more spots in the equatorial zone and belts; he was very much a product of the European tradition of rotation-timing. Lassell's spots, which appeared in Flammarion's spring 1875 drawings, were the features nearest the pole with which he concerned himself [28].

While Brett was watching the equatorial zone, Joseph Gledhill was following three bright spots in the SPR. Most well documented features until now had been in the jovian mid-latitudes; thus, Gledhill's spots, while typical in morphology, were exceptional in location. 'They seemed quite round, about the size of Sat. I. [Io]', said Gledhill, 'when fairly within the disc ... and quite as bright as any of the bright parts of the surface of the planet'[29]. They remained in a 'dusky band' that may actually have been a south south temperate belt. Their proper motion was indicated by their changing relative position with respect to another feature, a 'gap' in this band. Gledhill had seen similar spots in 1869.

Other lesser known astronomers followed Parsons' and Bredikhin's precedent of publishing a series of drawings made over a period of time and referencing them to a standard diagram. This began to replace the practice of publishing drawings of only unusual or conspicuous features.

Edward Knobel compiled his Jupiter drawings for 1874 in this way[30]: all thirty-five were made at an $8\frac{1}{2}$ inch Browning reflector with achromatic eyepieces yielding 144×, 208× and 250×. Knobel also used a crude seeing scale to evaluate numerically the conditions under which he made each of his drawings. The 1874 apparition received, by-and-large, competitive marks.

Knobel was interested mainly in changes in albedo, band widths and the number of visible belts and zones. His nomenclature was particularly easy to follow but still used terms like 'south tropical belt' that are anachronistic today. Knobel was one of the earliest to write of the components of the NEB and SEB. He also documented an SEB disturbance and a recurrence of 'Dawes' (Lassell's) spots[31].

The colours on Jupiter surprised Knobel, and other observers in 1874, by becoming vivid again. The equatorial zone was once more 'bronze yellow or sienna'[32]. Flammarion expended many 'colourful' words summarizing the 1874 apparition:

> En 1874, à travers les diversités de chaque jour, on remarque un fait général: c'est que deux bandes s'étendent sur la région équatoriale, et que l'une, celle du nord [NEB], est *jaune* et *très-claire*, tandis que l'autre, celle du sud [SEB], est très-foncée et de *couleur marron* ou *chocolat* ... Un autre fait remarquable a été la différence de nuances des deux callottes polaires: la callotte boréale [NPR] a toujours été nuancée de bleu violacé, tandis que la calotte australe [SPR] est restée plus jaune et moins foncée. [In 1874, through the diversity of the days, one notices a general fact: two bands extend on the equatorial region and the northern one (NEB), is yellow and very

pale, while the other one, the southern one (SEB), is very dark and brown or chocolate . . . Another extraordinary fact was the difference of shades of the two polar segments: the boreal segment (NPR) has always been finely shaded of purplish blue, while the southern segment (SPR) has remained more or less yellow and less dark][33].

8.4   Published Drawings: From Few to Many

At this point, it may be useful to draw attention to a noteworthy change in the way observations of Jupiter were presented in publications of the day. Perhaps as another application of the neo-Baconian philosophy, one sees an increased tendency to publish drawings, by themselves and without comment, presumably as a way of presenting (or it was hoped) a more impartial account of what was visible on Jupiter.

One of the first series of such drawings had already been published by John Birmingham as early as 1871[34]. (By 'series' I mean a collection of drawings, closely spaced in time, made with the intention of establishing such a data base.) In part, it may be that the appearance of such series owes itself to the publishing industry's capacity and willingness to print a large number of drawings on multiple plates; in part it may also reflect the increased appreciation among astronomers of the variability in the features of the jovian disc, and the longitudinal asymmetry of such features around the planet.

However, the most important factor in such representations is the growing conviction that the observer should not be the one to cull and interpret the data. Instead, she or he should hold nothing back; everything should be included, and allowed to stand on its scientific merit.

In practice, of course, astronomers did continue to omit drawings selectively from their published collections. They also continued to comment upon their artwork, if only to please editors uncomfortable with the idea of a nearly wordless paper. Certainly we should be glad that they did caption their drawings. These captions allow us to compare most directly the artist's view of his or her drawing to our idea of it today. Comparison between an astronomer's depiction and written commentary about the same view of Jupiter provides the only reference point by which to figure out the astronomer's relationship to his or her work.

Indeed, we could wish for more comment and more editing. This is true particularly for drawings that even their makers, when viewing them in necessarily resolution-reduced published form, must have seen had very little information content. (Yet, to be consistent, was that for them to say?) Too, we could ask for more interpretation, though I will discuss in a later chapter why many observers were uncomfortable with theoretical speculation. Still, presenting uninterpreted 'raw' data with a minimum of annotation implicitly gives each datum (i.e. drawing) equal weight, a disservice to posterity.

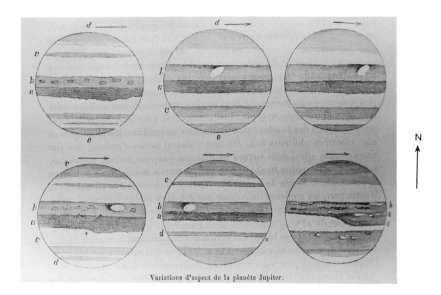

Variations d'aspect de la planète Jupiter.

Figure 8.5: Jupiter, 1875, by Flammarion. (From 1875 *Comptes Rendus Hebdomadaires des Séances L'académie des Sciences.* **81** 888; produced by Aaron Spurr.)

8.5 Possible South Temperate Ovals

One more prominent phenomenon remains to be documented. Major spots in the south temperate belt are difficult features with which to establish continuity between the 19th century and today. The reason for this is that the most familiar (and only) examples now, the anticyclonic White Ovals at $-34°$ latitude, have been in continuous existence for 50 years†. In other words, we have seen this set of White Ovals appear, but we have never seen them disappear.

The meagre evidence for earlier similar spots begins in 1872. In a drawing of 5 February of that year, by the Potsdam Observatory's Oswald Lohse (1845–1915), a single oval feature appears[35]. It is light with a dark annulus. (A low albedo 'collar' is characteristic in observations of both the White Ovals and Great Red Spot today.) The spot is too far south to be likely to be the GRS even if that feature had been visible inside the Hollow at the time. Instead, it is easier to imagine that it is akin to the White Ovals or, perhaps, an isolated and large example of Lassell's spots.

On 27 March 1873, in a Trouvelot drawing, the STB was dark except for a long oval in the western hemisphere[36]. It is reminiscent of the early evolution

† The story of the White Ovals is unique. Elmer Reese described the phenomenon as *dark* features appearing in a *light* zone. As the low-albedo features expanded, the high-albedo longitudes shrank and became the White Ovals.

of a White Oval. Trouvelot drew the same feature on 15 April.

It was only Knobel, though, who drew south temperate spots with any regularity[37]. He observed possible White Oval precursors in 1874, from March to June. However, while Knobel drew STB spots plentifully, most were apparently vague in nature, and none are rendered as definitively as Lohse's of 1872. Not even a hint of these features was published during the rest of the apparitions of the mid-1870s.

It is ironic. Astronomers the world over—who had been taken by surprise by the seeming major changes on Jupiter during the early 1870s—were now ready and waiting at their telescopes for a sequel of activity, but there was relatively little to observe†. (The fact that most of the world's astronomers were in the northern hemisphere, while Jupiter was now in the south celestial hemisphere, did not help the situation.)

The Great Red Spot Hollow continued to appear in artwork without commentary for two apparitions. It is a readily identifiable bulge in a drawing made by Pietro Tacchini on 28 January 1873[38]. It is at the limb in the Trouvelot drawing of 6 March[39]. The Parsonstown lithographs show it only once (dated 10 April)[40]. Knobel portrayed the Hollow at the beginning (7 March) of the 1874 apparition[41], but it then remained undocumented in 1875.

Scientists had conspired to attempt to solve the mystery that was the physical nature of Jupiter, but already the extent of the observational data seemed bounded and at hand. Was there nothing more under Jupiter's sun? A misleading halcyon air seemed to persist over the giant planet, as more eyes watched its disc than ever before. No one knew that this calm was about to be broken by the spectacular and unique appearance of the Great Red Spot.

8.6   Reports from the Southern Hemisphere

The best report of the 1875 apparition comes to us from 'down under' in New Zealand, which experienced particularly good weather during the first half of that year. We know the observer only as 'Miss Hirst', because an intermediary, 'S J Lambert', transferred her reports to the pages of the *Monthly Notices*[42].

Hirst's equipment was very similar to that of Knobel in England. She used an $8\frac{1}{2}$ inch Browning telescope and achromatic eyepieces with focal lengths for obtaining 144×, 208× and 250×[43]. Evidently, this was a standard telescopic system marketed by Browning.

Lambert's nomenclature was almost identical with the current one. Terms such as 'north temperate belt' were defined as they are today[44]. A phrase such

---

† Lohse and Flammarion depicted an SEB disturbance in 1875, but it cannot be said whether or not this was a continuation of the phenomenon from the previous year. (Lohse O 1878 Beobachtungen und Untersuchungen über die physische beoschaffenheit des Jupiter und Beobachtungen des Planeten Mars *Publicationen des Astrophysikalischen Observatoriums zu Potsdam* **1** 93; Flammarion C 1875 Observations de la planéte Jupiter *Comptes Rendus Hebdomadaires des Seances de L'academie des Sciences* **81** 887.)

as 'the equator' had its present meaning, being confined to those equator-straddling latitudes between the SEB and NEB. Lambert even mentioned Hirst's seeing the 'equatorial band'[45].

Reports on Jupiter relayed to the RAS from New Zealand may not seem so peculiar if we recall that, during this time, Jupiter was at a southern declination. The planet now reached altitudes in the sky necessary for optimal seeing only south of the equator. The northern counterpart to Hirst's description of the 1875 apparition was a series of rather crude chalk drawings produced at the United States Naval Observatory. The Superintendent of the Observatory readily admitted that observing conditions were poor in the District of Columbia during that year. He also made a few written remarks about the planet—for instance, a very nautical aside that the equatorial zone was 'banded in ... strands ... like a rope'[46].

In March, 1876, the Royal Astronomical Society issued a circular that called for a detailed study of Jupiter to search for periodicities in phenomena there[47]. Lohse had suggested as early as 1873 that northern hemisphere astronomers take up such a task. Now that Jupiter's declination was dipping below the celestial equator, the RAS called upon southern hemisphere observers to take up the watch.

The RAS committee that solicited these observations made up the first group organized specifically for the study of Jupiter. These persons, the signatories of the circular, represented a roll call of observers who as individuals had led, and would continue to lead, the study of the giant planet in the late 19th century. They were William Huggins, Knobel, Lord Lindsay†, Lohse, Arthur Ranyard, Lord Rosse, Terby and Thomas Webb.

Sydney merchant (and yachtsman) George Hirst (1846–1915)[48] (not, apparently, related to the 'Miss Hirst' mentioned above) had recently come upon the use of a $10\frac{1}{2}$ inch silvered Newtonian and decided to respond to the call of the Society committee[49]. He first looked back at the last 10–15 years of drawings of the planet to see what his predecessors had done. Hirst was surprised by the paucity of good drawings available. In his survey he was restricted to *published* artwork. An Australian, Hirst lacked access to those renderings periodically exhibited at RAS meetings and privately circulated in Great Britain and the rest of Europe. Indeed, as we have seen, those drawings (which found their way into books or journals that he might have had an opportunity to see) had been, until very recently, quite limited.

Hirst determined to improve both the quantity and quality of published pictures of Jupiter:

> ... I must confess it a matter of great surprise, that so few and such crude attempts have yet been made to give the general astronomical reading public an idea of the telescopic appearance of this, the most magnificent of our planets; and the reason I am at a loss to see; for

† Alexander Lindsay (Earl of Crawford; 1812–1880), the only unfamiliar name on this list, operated a private observatory at Dun Echt, Scotland.

Figure 8.6: George Hirst. (Courtesy of James Moors.)

as I have before said, Jupiter is certainly, excepting our Moon, the
easiest of all telescopic objects, and after a little practice, any one I
am sure, with a decent notion of using his pencil or chalks, may give
a far more accurate representation of the planet than he will find in
the most elaborate and expensive astronomical work he can lay his
hands on. Very few drawings ever represent colours at all; in a very
extensively got up work I have in my library the belts are represented
as straight lines—as if, to save trouble, they had been drawn with a
ruler; in others there is an attempt at a ragged, cloudy appearance,
but the artists who represented them evidently drew from what they
had heard rather than what they had seen. Messrs. De La Rue and
Lassell have both furnished what have been said to be remarkably
fine drawings, and probably the originals may be; but if this is the
case a lithographic copy of one of them that I have seen must be a
most woeful libel[50].

Hirst complained about Laurence Parsons' 1873 drawings made with the

6 foot aperture reflector: '...I have repeatedly observed more detail in the $10\frac{1}{4}$-inch reflector on an ordinary night than is shown in any one of them'[51]. Calling attention to the general redness that seemed to pervade the drawings, he suggested that this was due to the properities of Rosse's speculum metal mirror, and had previously invoked the same explanation to account for the overabundance of red stars in William Herschel's catalogue. (The reflectivity of speculum metal is directly proportional to wavelength[52].)

Now Hirst set about observing Jupiter himself. He found it a somewhat harder task than he had imagined.

The weather and sky conditions when Hirst began his observing were not good. He made up for this by being aware of and using the moments of good seeing that could flicker into existence on even a marginal night. He quickly discovered, though, that while his attentiveness was often rewarded by this kind of momentary glimpse of high resolution, making a drawing in this manner was difficult. Good drawings require a steady and patient model. Jupiter was neither. Often Hirst was frustrated by the impression that he had seen *something* extraordinary but was totally incapable of recording it:

> I have at times—but only, as I said before, for a second—seen the whole of the disc of Jupiter covered with fine lines; even the white belts, which ordinarily present not a trace of marking, are scored by them all over, and the darker equatorial zone appears a mass of flocculent, cloudy matter; but to attempt to put this on paper during the fleeting moment it is visible is an impossibility[53].

Hirst also had trouble with colour. He found that usually he had to use his full aperture to detect jovian hues at all. During an 1876 reappearance of the red equatorial zone, Hirst saw an 'orange-yellow' zone[54]. At the same time, his colleague, Henry Russell (1836–1907), observing with an 11 $\frac{1}{3}$-inch refractor, saw it as 'bright rose pink'[55]. The two men debated the 'authentic' colour until Hirst saw the same pink hue in a smaller refractor (though it was not as pronounced at lesser apertures). Meanwhile, Russell (Director of the Sydney Observatory) built an 11 inch silvered reflector, and he too saw the zone as orange-yellow.

This discrepancy puzzled Hirst. He reasoned thus: the refractor, even with the best achromatic lenses, still gave a purplish border to any bright disc such as Jupiter. It should stimulate the eye to see the complementary colour, yellow. Conversely, the reflector ought to add a slight red cast from a property of the silver. (In fact, silver is only slightly more reflective at longer optical wavelengths[56]). So, Hirst thought, the opposite colour biases from what was being observed should be produced.

A cryptic comment by Hirst now catches our attention as we approach the eve of the modern discovery of the Great Red Spot:

> Of the markings generally on the planet there are one or two which I will mention as being particularly characteristic and persistent. The

strangest-looking of them is the one Mr. Russell and myself called
the 'fish', on account of its presenting something of that form[57].

This observation dates from 24 May 1876. Could it have been of the Great
Red Spot? Ironically, Hirst, who complained about a lack of published jovian
drawings, did not at the time publish a drawing of this feature. Like so many
of his colleagues in Great Britain, he simply referred to a drawing presented
at a meeting of the local learned society (here, the Royal Society of New
South Wales). Furthermore, Hirst's written description[58] did not mention
the *location* of the belt where the feature occurred. Context only tells us that
it was one near the equatorial zone.

Two years later, after the modern discovery of the Great Red Spot, Hirst
wrote about the GRS, *alias* 'the Fish ... being on the same side of the equatorial belt, the only difference being that in that year [1876] it was incorporated
in the belt, from which it is at present [1878] well detached; its shape is exactly
the same, though it is now reversed, the preceding end tapering off, instead
of the following'[59].

Elsewhere, Hirst described the 1876 feature as 'bright red'[60]. As we have
seen, he normally had trouble with planetary colour. That Hirst was moved
to use this language suggests that the feature was truly extraordinary. (Re-
alistically, today *nothing* on Jupiter—including the Great Red Spot—can ob-
jectively be described as 'bright red'.)

When Hirst's drawing finally did make it to the RAS, in late 1877, Browning wrote to Hirst and said:

> Your drawing [of Jupiter] is the best I have seen for a long time. I
> will have it framed and exhibit it next month at the first meeting of
> the Royal Astronomical Society[61].

Hirst eventually became a member of the British Astronomical Association
and was elected a member of the RAS. His latitude put him in position to
observe Venus in transit as well as southern double stars[62].

Perhaps it is for the best that precedence does not honour Hirst with the
discovery of the Great Red Spot. Otherwise, we might yet today see Galileo
spaceprobe and Hubble Space Telescope scientists captioning high-resolution
images of the 'Great Red *Fish*'!

8.7   The Great Red Spot

Little happened in 1877. Sir Charles Todd (1826–1910), another Australian
(Adelaide Observatory), thought that Jupiter's southern hemisphere in that
year was not so coloured as in 1876. The story, such as it was, could be found
in the dark NEB. The belt appeared 'foggy'[63].

In November 1878, after searching for some mention in the astronomical
journals of what he had seen, Carr Pritchett (1837–1888) decided to write to

Figure 8.7: Pritchett's telescope. (From 1936 *Popular Astronomy* **44** 480.)

the editor of the new *Observatory* himself[64]. He described the feature he first observed on Jupiter, beginning the night of 6 July 1878, as an

> ...elliptic cloud-like mass, separate from the general contour of the belts. This cloud was almost a perfect oval in shape, and was preeminently rose-tinted. But the most remarkable phase of all was the rapid *proper motion* of this 'elliptical cloud'[65].

Pritchett (Morrison Observatory)† had unequivocally seen what we now know as the Great Red Spot. Three data of evidence present themselves in this single quotation: the morphology of the feature, its colour, and, perhaps most importantly (as we shall see), its motion with respect to other disc features. Never before had an observation met all three criteria. The drawing that Pritchett included with his letter is conclusive. It is of no vague detail, but of a definite huge spot, disrupting the parallel jovian belt/zone pattern in the southern hemisphere[66]. Only Pritchett's approximate latitude was questionable[67]; he quickly refined his '40°' to the current value[68].

While Pritchett's dated drawing[69] gave him precedence, the 'new' feature was seen almost simultaneously by observers stationed among the planetary science community, from Europe to Brazil[70]. An observer in India belatedly

† Fayette, MO, USA.

Figure 8.8: Jupiter, 9 July 1878, by Pritchett. (From 1879 *Observatory* **2** 308; courtesy of the Royal Astronomical Society.)

reported to the RAS his 'discovery' of an 'elongated oval form' in 1880![71]

(Even today, texts variously give Pritchett, Ernst Tempel (1821–1889), Louis Niesten (see below) or even Gledhill 'credit' for the modern discovery of the Great Red Spot. There seems to be some dependence upon whether the source was written in the United States, Germany, Belgium or the United Kingdom ...)

Those who did not 'discover' the Great Red Spot, 'recovered' it from their own drawings or observing notes. (These include Birmingham, Bredikhin, Knobel and Terby[72].) The Reverend James Virtue claimed to have seen the 'large detached red spot' just ten days before Pritchett[73]. Unfortunately, the entry he made, in the observing diary at the Dun Echt observatory that night, does not record it. No one questioned Virtue.

Alternately, observers quickly spotted the Red Spot when the first reports from others reached them. An example is Miklós Konkoly Thege (1842–1916) who drew it from his country estate in Hungary[74]. A wave of excitement propagated through devoted Jupiter-watchers and other astronomers alike as the greatest jovian feature of all was recognized. The number of its telescopic admirers— using apertures as small as $4\frac{1}{2}$ inches[75]—quickly reached hundreds. Birmingham announced the GRS to the world (*via* the pages of *Nature*) as a 'strange and beautiful feature like a flame–red elliptical cloud surrounded by brilliant white aureole' south of the SEB[76].

Commencing on 25 September 1878, Trouvelot observed the Great Red

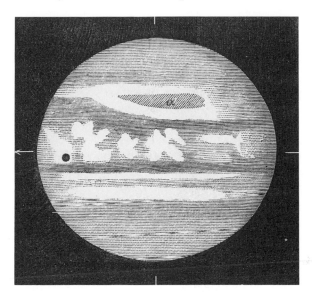

N

Figure 8.9: Jupiter, 25 September 1878, by Trouvelot. (From 1879 *Observatory* **2** 411; courtesy of the Royal Astronomical Society.)

Spot until 30 December, when Jupiter disappeared into twilight†. He described it, referring repeatedly to its hue:

> ... a very remarkable red spot was seen just a little above the southern edge of the equatorial belt ... This curious object, which apparently occupied one fifth of the planet's diameter, was very conspicuous, its intense rose-colour appearing in strong contrast with the white luminous background on which it was projected. It was of the same uniform shade throughout, without any dark border, its vivid rose-colour forming the whole spot. It appeared isolated from and perfectly independent of the equatorial belt, from which it was separated by a brilliant white band. In shade, the colour of this spot differed totally from the pale pinkish colour of the equatorial belt, or from any thing I have seen on Jupiter; a mixture of vermillion and white would very nearly give the shade of this mark[77].

After reading of Pritchett's discovery, Trouvelot questioned whether the two astronomers were discussing the same feature. Why had not Pritchett observed the feature again, later in July? Why had Trouvelot not seen it earlier in September, though he had been observing Jupiter (at the proper longitude on the planet) all month? He could not have missed such a glorious feature! Trouvelot's conclusion: no spot existed during the first three weeks

† By 1881, Trouvelot had made 567 drawings of Jupiter! (Trouvelot E 1881 Observations on Jupiter *Proceedings of the American Academy of Arts and Sciences* **16** 299.)

of September; Pritchett's 'oval' had died. *Two* phenomenal spots had formed, at different times, but at the same longitude! He found this coincidence 're-markable', but not beyond his experience with other recurring spots[78]. Some 'local causes' must precipitate impressive spots there, thought Trouvelot[79]. Moreover, he proposed that spots like Pritchett's (or Trouvelot's 'own' spot, should it survive conjunction) would be good jovian rotation-period indicators. Chained to a certain longitude, they would not exhibit proper motions of their own.

Pritchett had not observed the Great Red Spot between 18 July and 12 August for an ironic but straightforward reason: He was away from his $12\frac{1}{4}$-inch Clark refractor, on an expedition to watch the Great Solar Eclipse of 1878. What puzzled Pritchett was that he did not see the spot *afterward*, when observing Jupiter immediately upon his return[80].

Most observers believed there was but *one* Great Red Spot[81]. Louis Ni-esten (1844–1920) of the Royal Observatory (Brussels)[82] had decided to sketch Jupiter on every clear night between 16 September and the end of October 1878. (He used a 6 inch equatorial at 270×[83].) When he read of Trouvelot's observations, Niesten checked his sketch book. There was the GRS, in a drawing made on 6 August. Why had he not spotted it at the time? Unlike Pritchett and Trouvelot, who had happened to catch the Spot on the planet's central meridian, Niesten observed when the Red Spot was near the limb and less obviously a new feature.

Pritchett's, Trouvelot's and Niesten's spots were all at the same longitude. Niesten only could take so many coincidences. He considered Pritchett's July observations, his own of August and September and Trouvelot's in December to be a continous record of a single Great Red Spot[84].

And it was big. Bredikhin (whose 1876 observations show a hint of a GRS Hollow[85]) used his micrometer in 1879 to show that the Great Red Spot extended 16 arc seconds in longitude, covered 4 arc seconds in latitude and was situated 9 arc seconds south of Jupiter's equator[86]. (The equatorial diameter of the planet was 45 arc seconds.) Niesten put the GRS at 13 arc seconds by 3 arc seconds[87], while Professor George Hough (Dearborn Observatory†; 1836–1909) got 12 arc seconds by 4 arc seconds[88]. Trouvelot thought the Great Spot was becoming shorter but wider (extending further south)[89]. Regardless, it spanned about ten degrees of latitude and 30 degrees of longitude. This means that the Spot was roughly 20 000 km long and 4 000 km wide. Was it a brand new belt in the act of materializing?

From the 1880 *Scientific American*:

> The great red, elliptical spot on the visible surface of Jupiter is so long that could the earth be placed at one end of it and rolled it would make nearly a complete revolution before arriving at the opposite end; and so wide at the widest part that the earth would overreach it on either side by but little more than half the diameter of our

† Chicago, IL, USA.

moon, and stands in such contrast to the surrounding disc as to be visible with large telescopes when the planet is but three hours from the Sun in right ascension, and the Sun on the meridian[90].

Frank Dennett set a record for observing the 'oval pinkish spot' during the 1878 and 1879 apparitions[91]; he saw it on no less than 42 evenings using a $5\frac{1}{2}$-inch reflector. Dennett was first to comment on what soon would be most apparent: that the Great Red Spot was exceptional, if only for the different ways different people saw it.

Dennett already had seen its hue described as 'rose- tinted, fiery copper-coloured, red, vermillion, blood- red, and reddish or orange brown'[92] ('Indian red' and 'Venetian red' would pop up later[93], as would 'glowing pinky red'[94]) In German it was 'grauroth'[95] in French 'rouge très vive' [very bright red][96]. The *Monthly Notices of the Royal Astronomical Society* translated a Hungarian description as 'dirty flesh colour'[97]! Dennett felt that some of these differences were due to instrumentation. Others were due to real *changes* in the Spot. Still others were the result of a personal 'chromatic equation', unique to each observer[98]. (Dennett himself wrote for the record: 'decidedly *red*, with perhaps a very tiny tinge of yellow'[99].)

A few observers (for example, Webb) considered the Great Red Spot to be paler in its centre, when compared to its borders[100]. One thought the colour deeper along the major axis[101]. Dennett warned that the 'aurcole' observers sometimes thought they saw surrounding the Spot, and other imagined inhomogeneities, might be merely effects of contrast[102].

Dennett was the first to mention 'trails' of materials, either preceding or following the Great Red Spot. Sometimes they were grey; other times they were the colour of the GRS. Not one to be distracted by a pretty oval, Dennett also reported on white spots in the equatorial zone[103]. (Gledhill saw 'small detached' spots preceeding the Great Red one[104].)

Niesten and Captain William Noble (1828–1904) began to link the Spot with the Hollow[105]. Noble wrote of the 'curious assimilation of the contour of the surroundings of the red spot to the outline of the spot itself'[106]. Something physical was distorting the parallel band pattern. Noble believed that he saw preceding shadows (*à la* Brett) cast by it. A spherical globule would produce a shadow; however, Noble also wrote of a second shadow, following the Great Red Spot, for which he lacked an explanation. Later, Lohse pointed out that the GRS lost intensity and colour when near the limb[107]. This was paradoxical behaviour for a feature supposed to project far enough above the jovian clouds to cast observable 'shadows'[108].

Why the sudden availability of detail in the STrZ/SEB? BAA Jupiter Section Director John Rogers reminds us that the SEB faded to indistinguishability in 1879[109]. The solid white STrZ/SEB provided a striking canvas, on which the contrasting Great Red Spot was 'painted', until 1882.

What of the Red Spot's vivid colour *circa* 1880? Brett had a score to settle on this point:

Figure 8.10: Jupiter, 4 October 1879, by Noble. (From 1880 *Monthly Notices of the Royal Astronomical Society* **40** 87; courtesy of the Royal Astronomical Society)

Writers on astronomy are usually rather wild in their descriptions of phenomena, and their illustrations are often ridiculous.

There is at present a very extraordinary phenomenon in the solar system, more extraordinary, to my judgment, than any thing I have ever had the good fortune to see during about 30 years of telescopic practice; and it is thus described by writers in two scientific periodicals supposed to be serious:—

One of them says it is 'a beautiful feature like a flame'[Birmingham]. This occurs in the second line of his description. He cannot get even through two lines without forsaking the facts and telling you the thing is like something else, which I need hardly say, it is not. You, however, not having seen it go away with a definite and totally misleading image in your mind, and expect at all events to see a bright and coruscating appearance. Most observers have seen it by this time; but for the benefit of those who have not I may as well say at once that the most astonishing characteristic of this 'beautiful feature like a flame' is its dense darkness.

What does the other writer say? He sets out by calling it in his first line a 'beautiful cloud.' If he had said a 'beautiful coalscuttle' it would have conveyed about as exact an idea.

The word cloud conveys, to most minds, the image of something

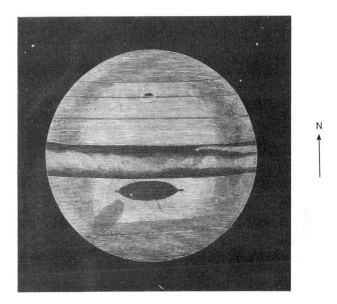

N

Figure 8.11: Jupiter, 25 July 1880, by Barnard. (From 1880 *Scientific American* **43** 356.)

vaporous, transient, and indefinite. Surely the study of astronomy must have a fatal tendency to make men vague! Allow me to say that if you cut a bit of maroon velvet into an elliptical shape, about an inch long, and stick it on to an orange you will see something quite as like a cloud as the phenomenon in question, and a good deal more like the thing itself[110].

How did Brett describe the new 'patch'?

... to call it blood-colour without qualification would be an exaggeration; but a drop of blood thickened by standing matches it pretty well. I speak of it as seen in a silvered-glass reflector. In a refractor I cannot say what colour it might not show; but I should be surprised to see it of a rose-colour, as it is often said to be. Surely it is not more like a rose than a geranium or a magpie![111]

Agreement on colour was important because colour suggested mechanism. 'Red' equalled 'hot' in the minds of some observers. Perhaps the great cloud (if it were a cloud) was simply reflecting the light produced by a tremendous lava flow on the 'surface' below ... Why the seemingly sudden breach in the jovian crust? Maybe it was due to the impact of a small body with Jupiter—an asteroid or comet. While the 'crust' of Jupiter has long since been forgotten, the idea that the Great Red Spot might be the result of cosmic collision has received reconsideration in light of the Shoemaker-Levy 9 impacts.

N

Figure 8.12: Jupiter, 3 September 1879, photographed by Common. The Saturn image is from 1885. (From the frontispiece of Clerke A 1908 *A Popular History of Astronomy during the Nineteenth Century* (London: Black); courtesy of David DeVorkin (Smithsonian Institution).)

The Reverend Edmund Ledger (1841–1913) thought just the opposite. To him, the Great Red Spot looked like a solid crust *forming*: the appearance of a new jovian continent[112].

An even more imaginative 19th-century suggestion for the genesis of the Red Spot was that Jupiter was about to give birth to a new satellite! Current theory suggested that satellites 'spun off' their parent planet; the early GRS was interpreted as such a fission in process[113].

(Interestingly, the actual phrase 'Great Red Spot' does not seem to have been coined by any regular observer of Jupiter. It first appeared in a popular London magazine, but not until 1880[114].)

Announcements of Great Red Spot 'pre-discoveries' appeared soon after the 1878 apparition[115]. Most notable among these was Hirst's (though Dennett, who had been observing in 1876, found it 'hard to believe that such an object as the red spot could have been missed'[116].) Ledger compared the GRS to a spot observed by Laurence Parsons (and mentioned earlier in this chapter) in 1873[117]. (It was at approximately the right latitude, and Parsons spoke of 'red'.)

Gledhill had observed a southern feature in 1869, and again in 1871; others saw it yet in 1872. (This was Alfred Mayer's 'Ellipse'.) A 'dark spot of elliptical form' was seen in 1868[118]. The 1878 appearance and observations of the Great Red Spot caused Albert Holden (1848–1897) to ask whether great spots were co-periodic with the sunspot cycle[119]. (See chapter 9.)

Piazzi Smyth played scholarly detective with the observing notebooks of the recently deceased Reverend Henry Key (1819–1879)[120]. Key had seen an oval the size and shape of the Great Red Spot, but he had labelled it as black. Smyth pointed out that in Key's *4-inch* aperture speculum, perception of colour was not to be expected. The date for Key's oval was 4 June 1843.

Others looked further back into history. The Reverend Samuel Johnson (1845–1905) recalled Johann Schröter's 1792 spot[121]. Brett (and many others after him) wondered if there was any correspondence to the 1665 spot of Cassini[122]. (See chapter 10.) If the Great Red Spot was a 'lava flow', it was an improbably long-lived one.

Still, how long would this spectacle last? Few spots had been observed for more than two apparitions. However, there was no precedent for a spot covering an area equal to three-quarters that of the Earth's surface! Was the Great Red Spot something permanent?†

8.8   The Rotation of Jupiter

Meanwhile, Pritchett in America reminded observers in Europe that his spot was not staying put. It kept in step with small 'triangular spots' just south of it, but translated (at a constant rate) with respect to the white spots in the equatorial zone[123]. Based on the current popular estimate for Jupiter's rotation period (Albert Marth's (1828–1897) 9 h 50 min 8 s[124]), the Great Red Spot had drifted 21° to the west over the course of its first year.

Whether features on Jupiter were anchored somehow to the bulk of the planet below had yet to be resolved. Maybe, speculated Pritchett, some spots move, and some do not. Either proper motion is restricted to the equatorial zone, or the motion of the GRS is 'contrary' to that of the EZ[125]. Others complained that he had mismeasured his spots[126]. Pritchett held firm.

As recently as 1859, yet another observer had attempted to answer the 17th and 18th-century mystery of Jupiter's 'variable' rotation time[127]. Do different spots at different latitudes move at different speeds? Do the *same* spots move at different speeds at different *times*? If either of the above is true, do equatorial spots tend to move faster than more poleward ones? (This was Giovanni Cassini's conclusion.)

Based on multiple observations of five colatitudinal spots, Joseph Baxendell, too, came to believe that individual spots vary in their rotation periods and that these periods change with time. Moreover, he announced (somewhat surprisingly) that there was no correlation between spot rotation period and latitude[128]! However, he also warned:

† Telescope-maker Andrew Common (1841–1903) successfully photographed the GRS on 9 September 1879. (Photographs of Jupiter 1880 *Observatory* **3** 314.) He used a 36-inch aperture silvered reflector. (Clerke A 1908 *History of Astronomy During the Nineteenth Century* (London: Black); King H 1955 *History of the Telescope* (Cambridge: Sky).)

It is hardly necessary to remark that the results now given afford no certain information as to the period of rotation of the planet itself, as distinguished from that of its spots. On the contrary they seem to me to indicate very clearly—especially when taken in connexion with the results obtained by former observers—that in the present state of our knowledge of the phenomena which take place on the surface of the planet or in its atmosphere, any conclusions to its exact period of rotation based upon observations of the times of rotation of its spot must necessarily be very precarious[129].

Twenty years later, there was still resistance to a differential rotation rate. Furthermore, to many, the Great Red Spot offered an opportunity to establish Jupiter's rotation period definitively. There was one jovian rotation period and that of the GRS was it. Dr Henry Pratt timed the Spot for six months and found that it travelled around Jupiter in 9 h 55 min 33.91 s[130]!† Thomas Brewin (1840–1927) obtained 9 h 55 min 34.1 s[131].

However, in autumn of 1880, an unusually bright, white, equatorial spot (a snow-capped mountain peak?) exhibited a rotation period significantly shorter than that of the Great Red Spot. This was obvious for all to see as the spot overtook the GRS during the apparition. Indeed, it did so every 44 days[132].

The spot had been observed the year before—by Terby and Dennett—but no significance had been attached to it until William Denning (1848–1931) and others undertook to regularly time the new spot's rotation period[133], as had become the custom with the Great Red Spot. Indeed, Denning watched both it and the GRS for over a year, at 150×, commencing on 24 December 1881. (He, too, had a Browning reflector; its aperture was 10 inches.) Denning found that, while the Red Spot rotated around the jovian globe 967 times in this interval, the 'brilliant equatoreal spot' did so 976 times[134]. The Bristol accountant[135] produced an ephemeris of dates and times for when the two spots would be in 'conjunction' on the central meridian[136]. Owing to the 'irregularities' in the motions of each, some of these events occurred up to over a day late *or* early[137]. At most, only *one* of these spots could be a 'surface' feature.

An equatorial dark spot, discovered by Denning in October 1882, happened to have the same latitude as the white one. He timed it and obtained the same rotation period. This allowed Denning to put to rest an old controversy: '...we may assume that all the various markings both light and dark along the equator participate in [the same] swift movement ...'[138]. In other words, spots at a given latitude travel at the same speed, regardless of albedo (or size). This would be expected of clouds—but not of a mixture of clouds and 'ground' features (imagined below them). Furthermore, it did not matter in

---

† It is likely that Pratt had unknowingly seen the spot before. Back in 1872 he wrote of '...a dusky marking, its western end having an oval outline'. (Pratt H 1873 Jupiter *The Astronomical Register* **10** 42.)

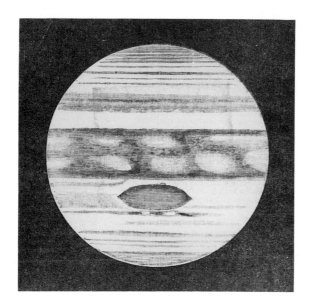

N

Figure 8.13: Jupiter, 7 December 1881, by Denning. (From 1882 *Nature* **25** 224.)

what band or albedo a spot travelled! From Denning we first hear of the latitude-specific 'currents' that would influence thinking about jovian spot behavior in the next century.

By 1883, Denning had identified consistently three separate rotation periods of Jupiter: that of the white EZ spot (9 h 50 min) that of the Great Red Spot (9 h 55 min 34 s) and that of 'dark, irregular spots' discovered October 1880 by Dennett in the northern jet (9 h 48 min)[139]. (The northern spot was the fastest so far ever recorded on Jupiter; it lapped the GRS every 32 days.) Hough had concluded that there were two rotation rate 'systems' at work on Jupiter, one in the tropics and one elsewhere, and that the tropical rate was a function of latitude. Denning now came to a different realization:

> It is evident that the different velocities of the various spots offer a very complex problem for solution. It may be that though these velocities follow no law dependent upon distance from the equator, they may each be special to the zones in which they occur[140]†

Both before and after the time of Wilhelm Beer and Johann Mädler, astronomers had attempted to measure a mean rotation period for Jupiter, but most of the measurements upon which these determinations were based used

† Denning's group of enthusiasts (later the Liverpool Astronomical Society) was a progenitor of the BAA, which took the lead in monitoring Jupiter during much of the 20th century. (Baum R 1998 The OAS—birth of a legend *Journal of the British Astronomical Association 108*)

more-equatorward spots than the Great Red Spot. Thus, the GRS gave the unacceptable appearance of moving in a retrograde direction!

Which period was 'real'? Was even the Great Spot blowing in surface winds? If so, it could be construed as translating—at a speed up to hundreds of kilometres per hour—in a direction *opposite* that of the rotating planet (so as to make the Spot 'tardy'). Yet, if one feature's motion was a vector sum of the planet's rotation and local wind speed, might not *all* jovian spots be driven by winds?

The periods of both the Great Red Spot and the bright 'Denning's Spot' were thought to have lengthened slightly by 1883[141]. Still, they remained obviously—and irreconcilably—different. (The GRS also had finally faded somewhat from its 3-year splendour in 1882, not to revive until 1886[142].)

Cassini, William Herschel, Schröter, and Airy had all been confronted with the possibility of differential rotation on Jupiter. It could no longer be denied. The coincidence of the dramatic Great Red Spot and a highly contrasting EZ spot appearing at the same time forced observers to conclude what they already suspected: there were (at least) *two* rotation periods of Jupiter. One applied in the equatorial regions, the other more poleward. This fact was to be codified into the System I and System II longitude coordinates used through much of the 20th-century. Denning was even more progressive:

> In any case, we may be assured that all the varieties of spots and streaks observed so persistently and numerously upon the planet are the phenomena of his dense atmosphere and not objects of stability upon his actual surface, which seems to be entirely shrouded from our scrutiny .... The further telescopic study of these remarkable markings cannot fail to prove of considerable interest, and may perhaps lead to the partial elucidation of what now appears to present difficulties beyond explanation[143].

(William Denning received the RAS Gold Medal in 1898, but not for his planetary observations: It was he who discovered the radiants of meteor showers[144].)

8.9   Summary

I have dwelt on the observing careers of Laurence Parsons, Fedor Bredikhin† and several other planetary astronomers of the mid-1870s to explain the maturation of observing methodology. One feature of this process was the explosion of Jupiter drawings made during these years. Another was the linkage of individual observers into a nearly world-wide network for the first time. This

---

† As late as 1880, Bredikhin tried to reconcile Jupiter's seeming dual rotation periods by postulating an equatorial zone at a much greater elevation than the rest of the planet. (Bredikhin F 1881 Sur la constitution de Jupiter *Astronomische Nachrichten* **99** 25.)

network was ready for the Great Red Spot and pounced upon that feature when it gave the first hint of improved visibility.

The Red Spot reached its greatest prominence, so far ever witnessed in 1879–81. It could be made out in simple opera glasses. The GRS was so big and so obvious that it provided an unambiguous marker by which to note the latitudinally dependent rotation periods of Jupiter. This fact could not be ignored any longer. The search for a *visible* solid 'surface' of the giant planet all but ended in the 1880s.

Endnotes

(All titles are written in full, with the exception of '*Mon. Not.*' for the *Monthly Notices of the Royal Astronomical Society*.)

1. Airy G 1849 Substance of the lecture delivered by the Astronomer Royal on the large reflecting telescopes of the Earl of Rosse and Mr Lassell, at the last November meeting *Mon. Not.* **9** 110
2. Parson L 1874 Notes to accompany chromolithographs from drawings of the planet Jupiter, made with the six-foot reflector at Parsonstown, in the years 1872 and 1873 *Mon. Not.* **34** 235
3. *Ibid.*
4. *Ibid.*
5. *Ibid.*
6. *Ibid.*
7. *Ibid.*
8. Winlock J 1876 *Astronomical Engravings of the Moon, Planets, Etc.; Prepared at the Astronomical Observatory of Harvard College.* (Cambridge: Wilson)
9. Parsons L *Op. Cit.*
10. *Ibid.*
11. King H 1955 *The History of the Telescope* (Cambridge: Sky)
12. 1912 François Joseph Charles Terby *Mon. Not.* **72** 257
13. Terby F 1874 Extract from M Terby's letter dated Louvain, May 1, 1874. *Mon. Not.* **34** 402
14. Parsons L 1874 Note on drawings of Jupiter made at Parsonstown, and by M Terby, at Louvain, in 1873 *Mon. Not.* **34** 402
15. Bredikhin T 1874 Observations sur le Jupiter faites en 1874 *Bulletin de la Société Impériale des Naturalistes de Moscou* **47** 87
16. *Ibid.*; Bredikhin T 1876 Observations sur le Jupiter *Annales de l'Observatoire de Moscou* **2** 42; Bredikhin T 1877 Observations sur le Jupiter *Annales de l'Observatoire de Moscou* **3** 127; Bredichin T 1878 Observations sur le Jupiter *Annales de l'Observatoire de Moscou* **4** 77
17. Bredikhin T 1874 Observations sur le Jupiter faites en 1874 *Bulletin de la Société Impériale des Naturalistes de Moscou* **XLVII** 87
18. *Ibid.*

19. *Ibid.*
20. Bredikhin T 1877 Observations sur de Jupiter en 1876 *Annales de l'Observatoire de Moscou* **3** 127
21. Bredikhin T 1876 Observations sur de Jupiter *Annales de* l'Observatoire de Moscou **2** 42
22. Bredikhin T 1877 Observations sur de Jupiter en 1876 *Annales de l'Observatoire de Moscou* **3** 127
23. Bredikhin T 1878 Observations sur de Jupiter en 1877 *Annales de* l'Observatoire de Moscou **4** 77
24. Brett J 1874 Memorandum of observations of Jupiter made during the month of April 1874 *Mon. Not.* **34** 359
25. *Ibid.*
26. *Ibid.*
27. Flammarion C 1875 Observations del la planète Jupiter *Comptes Rendus Hebdomadaires des Séances de L'academie des Sciences* **81** 887
28. *Ibid.* and Flammarion C 1877 *Terres du Ciel* (Paris: Libraire Académique Didier)
29. Gledhill 1874 J Bright spots on Jupiter *Mon. Not.* **34** 360
30. Knobel E 1874 Observations of Jupiter, 1874 *Mon. Not.* **34** 403
31. *Ibid.*
32. *Ibid.*
33. Flammarion C 1875 Observations de la planète Jupiter *Comptes Rendus Hebdomadaires des Séances de L'academie des Sciences* **81** 887
34. Birmingham J 1871 Beobachtungen des Jupiter *Astronomische Nachrichten* **77** 301
35. Proctor R and Ranyard A 1892 *Old and New Astronomy* (London: Longmans, Green)
36. Winlock J *Op. Cit.*
37. Knobel E *Op. Cit.*
38. Tacchini P 1873 Sur quelques phénomènes particuliers offerts par la planète Jupiter pendant le mois de janvier 1873 *Comptes Rendus Hebdomadaires des Séances de L'académie des Sciences* **76** 473 Referring to the original painting, in 1873 La Planète Jupiter *La Nature* 357, Amédée Guillemin describes the feature this way: '. . . dont le contour bordé de blanc se projetait sur un fond de couleur rose'.
39. Winlock J *Op. Cit.*
40. Parson L 1874 Notes to accompany chromolithographs from drawings of the planet Jupiter, made with the six-foot reflector at Parsonstown, in the years 1872 and 1873 *Mon. Not.* **34** 235
41. Knobel E *Op. Cit.*
42. Lambert S 1875 Remarks on drawings of Jupiter made by Miss Hirst, at Auckland, New Zealand *Mon. Not.* **35** 401
43. *Ibid.*
44. *Ibid.*
45. *Ibid.*

46. Davis C 1876 Drawings of Mars and Jupiter made with the 26-inch equatoreal of the United States Naval Observatory *Mon. Not.* **36** 13

47. Hirst G 1877 Some notes on Jupiter during his opposition of 1876 *Journal and Proceedings of the Royal Society of New South Wales* **10** 83

48. Moors J 1997 personal communication

49. Hirst G *Op. Cit.*

50. *Ibid.*

51. *Ibid.*

52. Hodgman C (ed) 1957 *Handbook of Chemistry and Physics* (Cleveland: Chemical Rubber Company)

53. Hirst G *Op. Cit.*

54. *Ibid.*

55. *Ibid.*

56. Hodgman C *Op. Cit.*

57. Hirst G *Op. Cit.*

58. *Ibid.*

59. Hirst G 1879 Notes on Jupiter following his opposition, 1878 *Mon. Not.* **39** 238

60. Hirst G 1877 Some notes on Jupiter during his opposition of 1876 *Journal and Proceedings of the Royal Society of New South Wales* **10** 83

61. Haynes R, Haynes R, Malin D and McGee R 1996 *Explorers of the Southern Sky* (Cambridge: Cambridge University Press)

62. *Ibid.*

63. Todd C 1879 Observations at the Adelaide Observatory *Mon. Not.* **36** 1

64. Pritchett C 1879 Markings on Jupiter *Observatory* **2** 307

65. *Ibid.*

66. *Ibid.*

67. Backhouse T 1879 The Red Spot on Jupiter *Observatory* **3** 250

68. Pritchett C 1880 Motion of the Spots on Jupiter *Observatory* **3** 317

69. Pritchett C 1879 Markings on Jupiter *Observatory* **2** 307

70. See, e.g. Cruls Détermination de la durée de la rotation de la planète Jupiter *Comptes Rendus Hebdomadaires des Séances de L'académie des Séances* **91** 1049. Cruls was Director of the Rio de Janeiro Observatory.

71. Nursinga Row A 1881 Observations of the Red Spot on Jupiter *Mon. Not.* **41** 46

72. Denning W 1898 The Great Red Spot on Jupiter *Mon. Not.* **58** 488

73. Lindsay 1880 Note on the spectrum of the Red Spot on Jupiter *Mon. Not.* **40** 87

74. Illés-Almar E 1992 On Konkoly Thege's Jupiter Observations *The Role of Miklós Konkoly Thege in the History of Astronomy in Hungary* eds M Vargha, L Patkos and I Toth (Budapest: Konkoly Observatory)

75. See, e.g. Tebbutt J 1881 Observations of the red spot on Jupiter *Mon. Not.* **41** 331

76. Birmingham J 1879 The planet Jupiter *Nature* **20** 403

77. Trouvelot E 1879 On the recurrence of some of the markings on Jupiter *Observatory* **2** 410
78. *Ibid.*
79. *Ibid.*
80. Pritchett C 1879 The red spot on Jupiter *Observatory* **3** 174
81. See, e.g. Dennett F 1880 Red spots on Jupiter *Astronomical Register* **17** 117
82. Paquet P 1997 personal communication
83. Niesten L 1879 Note sur la tache rouge observée sur la planète Jupiter pendant les oppositions de 1878 et de 1879 *Bulletins de l'Académie Royale des Sciences, des Lettres et des Beaux-arts de Belgique* **48** 604
84. Slack H 1880 The red spots on Jupiter *Astronomical Register* **17**
85. Bredikhin T 1877 Observations sur de Jupiter en 1876 *Annales de l'Observatoire de Moscou* **3** 127
86. Bredikhin T 1879 Schreiben des Herrn Prof. Bredikhin an den Herausgeber *Astronomische Nachrichten* **95** 383
87. Slack H 1880 The Great Spot in Jupiter. *Astronomical Register* **17** 266
88. Hough G 1881 Jupiter *Astronomical Register* **18** 243
89. Trouvelot E *Op. Cit.*
90. Fitz H 1880 What is the time of Jupiter's rotation? *Scientific American* **42** 293
91. Dennett F 1879 Markings on Jupiter *Observatory* **2** 352; Dennett F 1879 The red spot or cloud on Jupiter *Observatory* **3** 206
92. Dennett F 1880 Spots on Jupiter *Observatory* **3** 318
93. Barnard E 1880 Changes on Jupiter *Scientific American* **43** 356
94. Calver G 1881 Jupiter *Observatory* **4** 21
95. Tempel W 1880 Schreiben des Herrn W Tempel an den Herausgeber *Astronomische Nachrichten* **46** 61. Tempel worked at various observatories in France and Italy.
96. Bredikhin T 1879 Schreiben des Herrn Prof. Bredikhin an den Herausgeber *Astronomische Nachrichten* **95** 383
97. de Gothard A 1883 Physical observations of Jupiter and Mars made during the year 1882 at the Astrophysical Observatory, Herény, Hungary *Mon. Not.* **43** 425
98. Dennett F *Op. Cit.*
99. *Ibid.*
100. *Ibid.*
101. McCance J 1881 Jupiter *Observatory* **4** 21
102. Dennett F *Op. Cit.*
103. *Ibid.*
104. Gledhill J 1880 The red cloud on Jupiter *Astronomical Register* **17** 236
105. Niesten L *Op. Cit.*
106. Noble W 1880 Note on two sketches of Jupiter *Mon. Not.* **40** 86
107. Lohse O 1880 Schreiben des Herrn Dr O Lohse an den Herausgeber *Astronomische Nachrichten* **96** 17

108. 1880 The Red Spot on Jupiter *Scientific American* **42** 293
109. Rogers J 1995 *The Giant Planet Jupiter* (Cambridge: Cambridge University Press)
110. .Brett J 1879 The great patch on Jupiter *Observatory* **3** 236
111. *Ibid.*
112. Ledger E 1882 *The Sun: Its Planets and their Satellites* (London: Edward Stanford)
113. Ley W 1966 *Watchers of the Skies* (New York: Viking)
114. Slack H 1880 What is Jupiter doing? *Belgravia* **40** 453
115. Holden A 1880 The great patch on Jupiter *Observatory* **3** 282
116. Dennett F 1881 Jupiter's satellite I and the red spot *Observatory* **4** 24
117. Ledger T 1880 The red spot on Jupiter *Observatory* **3** 449
118. Gledhill J 1880 Jupiter in 1869 and 1879—the 'Ellipse' and the 'Red Spot' *Observatory* **3** 279
119. Holden A *Op. Cit.*
120. Smyth P 1880 *Observatory* **3** 450
121. Johnson S 1880 Spot on Jupiter in 1792 *Observatory* **3** 283
122. Holden A *Op. Cit.*
123. Pritchett C 1880 Motion of the spots on Jupiter *Observatory* **3** 317
124. See, e.g. Marth A 1882 Addition to the ephemeris for physical observations of Jupiter *Mon. Not.* **42** 427
125. Pritchett C 1880 Schreiben des Herrn Pritchett, Directors der Morrison Sternwarte, an den Herausgeber *Astronomische Nachrichten* **96** 223
126. Backhouse T 1880 The red patch on Jupiter *Observatory* **3** 320
127. Baxendell J 1862 On the rotation of Jupiter *Memoirs of the Literary and Philosophical Society of Manchester* **1** 194
128. *Ibid.* The difference in mean rotation periods among spots exceeded the standard deviations of the rotation periods calculated for individual spots.
129. *Ibid.*
130. Pratt H 1880 On the rotation-period of Jupiter *Mon. Not.* **40** 153
131. Brewin T 1880 Rotation period of Jupiter *Mon. Not.* **40** 377
132. Denning W 1880 The markings on Jupiter *Observatory* **3** 654; Denning W 1882 The markings on Jupiter *Nature* **25** 223
133. Denning W 1881 The motions and varieties of the jovian spots. *Mon. Not.* **61** 44; Denning W 1881 Transit times of the spots on Jupiter *Mon. Not.* **61** 358; Denning W 1882 The relative motions of the Great Red Spot and brilliant equatorial spot on Jupiter *Mon. Not.* **42** 97
134. Denning W 1882 The relative motions of the Great Red Spot and brilliant equatorial spot on Jupiter *Mon. Not.* **42** 97
135. Rogers J 1995 *The Giant Planet Jupiter* (Cambridge: Cambridge University Press)
136. Denning W *Op. Cit.*
137. *Ibid.*; Denning W 1882 The markings on Jupiter *Mon. Not.* **43** 34
138. Denning W 1883 Observations of Jupiter *Mon. Not.* **43** 97

139. Denning W 1881 Observations of Jupiter *Observatory* **4** 83; Denning W 1883 The markings on Jupiter *Mon. Not.* **43** 34
140. Denning W 1883 The markings on Jupiter. *Mon. Not.* **43** 34
141. Denning W 1883 Physical observations of Jupiter *Mon. Not.* **43** 300
142. *Ibid.*; Clerke A 1908 *A Popular History of Astronomy During the Nineteenth Century* (London: Adam and Charles Black)
143. Denning W 1883 Observations of Jupiter *Mon. Not.* **43** 97
144. Moore P 1997 William Frederick Denning *Encyclopedia of Planetary Sciences* eds J Shirley and R Fairbridge (London: Chapman and Hall)

Chapter 9

Theories of Jupiter

*If [the jovian inhabitants'] Globe is divided like ours, between Sea and Land,
as it's evident it is (elfe whence could all thofe Vapors in Jupiter proceed?)
we have great reafon to allow them the Art of Navigation ...*

Christiaan Huygens, Fellow, Royal Society

1698

In Thomas Kuhn's book, *The Copernican Revolution*[1], the author argues that
scientific theory proceeds in a series of discrete revolutionary jumps. His
example is the acceptance of the heliocentric theory of the solar system, which
suddenly allowed science to see better its place in the cosmos. There was a
leap from Earth as centre of the Universe to Earth as planet. The observation
of planetary discs and satellites by Galileo Galilei affirmed this notion[2].

The recognition of the Earth as just one of many worlds swept away the
Aristotelian concept of the heavenly bodies as objects far removed from things
Earthly. By making the Earth an example of a planet, it subjected planetary
bodies to close-up inspection. One such body, the Earth itself, could be
scrutinized in exceptional detail. What was learned about the Earth could be
extrapolated to the other planets by induction. For the first time, it became
possible to attempt physical interpretation of the appearances of other worlds.
The paradigm (in Kuhn's terms) that these interpretations were based upon
initially was a planetary structure resembling that of the Earth: a dense
rocky body with a thin tenuous atmosphere, and, occasionally, a global ocean
surrounding it.

This planetary paradigm worked reasonably well for some members of
the solar system. As applied to Jupiter, it implied that the outer planets
should have many characteristics similar to the Earth, albeit on a much larger
scale. The paradigm seemed most useful in modelling Jupiter's clouds; it was
less so in endeavours to understand what lies behind the visible cloud deck.
This is readily demonstrable even today: laymen continue to have conceptual
difficulties with a planet where questions such as 'what is its surface like?'

have no physical meaning.

As astronomy, and particularly astrophysics, matured, to the thesis that Jupiter was a larger version of the Earth, there was added an antithesis that Jupiter was a smaller version of the Sun. Much of the observational history of Jupiter during the time under discussion is, from a conceptual point of view, the record of the partial establishment of this younger Sun-like reference for the planet and eventually the first attempts at a synthesis of the old and the new jovian models.

These attempts continue today. The current debate over the depth of the part of the jovian atmosphere that affects or controls what we see has its historical roots in the nineteenth century.

9.1   Initial Inquiries into the Physical Nature of Jupiter

As humankind began to traverse the globe in increasing numbers, and world-wide meteorology became more understood, knowledge gained about our own planet was 'tacked onto' the perceived models of Jupiter and the other planets. Information obtained by navigators on southerly voyages allowed William Herschel to conclude that the equatorial latitudes of Jupiter were regions where Earth-like tropical weather patterns occurred and that there existed jovian trade winds in the belts. Astrophotographer Dominique Arago (1786–1853) furthered the trade-wind idea but noted that the jovian pattern in some places seemingly must blow in the direction opposite to that on the Earth[3]. So far from the Sun, it was difficult to see an energy source for the jovian winds; asteroid-discoverer John Hind[4] (1823–1905) thought it might be heat generated (through friction?) by the rapid rotation of the planet[5].

Still, the physical processes at work on Jupiter were as obscure during the early 19th century as was the supposed 'surface' of the planet. The problem was with creating a model of the jovian atmosphere that mimicked the behaviour of the Earth's atmosphere. Jupiter's fast rotation could be supposed to account in some way for the apparently permanent nature of latitudinal atmospheric currents, which on the Earth are variable. (Or, as William Denning later put it, the bands were 'ldots masses of heated vapour which are woven into parallel bands by the effects of very rapid rotatory movement'[6].) More difficult to explain were long-lived storm-like disturbances, cyclonic and anticyclonic, which on the Earth appear and disappear quickly.

Even before the modern discovery of the Great Red Spot, disturbances had been observed frequently enough to conclude that individual ones last on Jupiter weeks or even months. The author of one article[7] spoke of two well defined spots in the southern component of the north equatorial belt that exhibited no change for a period greater than ten weeks. Equally puzzling, when considered with their seeming longevity, were the apparently sudden changes that affect jovian features after long periods of quiescence. The same

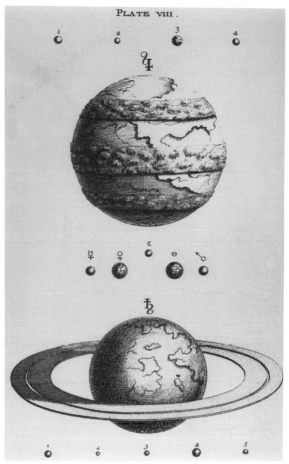

Figure 9.1: Thomas Wright's (1711–1786) theory of Jupiter. (From Wright T 1971 *An Original Theory or Hypothesis of the Universe 1750* (New York: Science History).)

article describes bright and dark NEB spots that vanished suddenly after several months.

In *The Planet Mars*, William Sheehan shows that the picture of an ever-changing Jupiter was so strong in the minds of some early observers that it caused them to remake other planets in Jupiter's image[8]. On Mars, semi-permanent surface albedo features *do* exist. Yet Giocomo Maraldi 'saw' cloud bands there, *á la* Jupiter. Similarly, interpretation of the white martian polar caps may have been impeded by the view that they were atmospheric in nature. (They are in reality standing icecaps.)

Early discussion about the solid 'surface' of *Jupiter* did not centre on its existence but rather on which feature represented it. Giovanni Cassini and

Christiaan Huygens believed that the dark belts were clouds and that the bright zones were ground[9]. They did not realize that, while the terrestrial clouds the men were accustomed to may in some circumstances look dark when viewed from below, they would be brightly illuminated and lighter than the ground if viewed from above. Johann Schröter made the same mistake[10]. Opinion gradually shifted to the view adopted by William Herschel: the zones were clouds, and the belts (and perhaps dark spots) were openings in them. However, both schemes continued to have their proponents. The debate became based on whether either the belts or zones seemed to disappear at the planetary limb†.

The notion of a solid 'surface' of the giant planet was still very much current over one-hundred years after Huygens. John Herschel:

> The parallelism of the belts to the equator of Jupiter, their occasional variations, and these appearances of spots seen upon them, render it extremely probable that they subsist in the atmosphere of the planet, forming tracts of comparatively clear sky, determined by currents analogous to our trade-winds, but of a much more steady and decided character, as might indeed be expected from the immense velocity of its rotation. That it is the comparatively darker body of the planet which appears in the belts is evident from this—that they do not come up in all their strength to the edge of the disk, but fade away gradually before they reach it[11].

At the same time the Great Red Spot was throwing Jupiter's differential rotation into the collective face of observers (*circa* 1880), there were still those who spoke of the jovian 'calotte polaire' [polar cap], to cite but one example[12]. Remarkably, it was not until the 1950s, when radio astronomers discovered the deep core rotation of Jupiter, that the idea of the planet's vast 'surface' entirely disappeared.

9.2   What is Heating Jupiter?

James Nasmyth (1808–1890) published the first attempt to detail the evolution of a planetary body such as Jupiter[13], though some of his ideas were anticipated by naturalist Georges Leclerc (Comte de Buffon; 1707–1788) and by philosopher Immanuel Kant (1724–1804)[14]. Nasmyth, an engineer, is known best for inventing the steam hammer (in 1839) and other machines. He retired with a fortune in 1856 and took up astronomy. With James Carpenter, he made initial investigations in selenology, the physics of the Moon.

It was with similar terrestrial-analogue planetophysics that Nasmyth approached the problem of Jupiter. His initial assumption was that the heat-

---

† Perspective causes the edges of both belts and zones to appear to converge at the limb. The 'lower' band should do so at a smaller angle from the central meridian (ignoring other phase effects).

retaining quality of a planet is governed by its mass, but that its heat-dissipating quality is governed by its surface area†. Thus, the giants of the solar system would take much longer to cool, assuming an initially hot formation. Nasmyth believed that a planet began in a molten state with a vaporized liquid and gaseous atmosphere enveloping it. Such was once the circumstance of the Earth. Nasmyth concluded that Jupiter (slower to lose heat, the principal evolutionary element in Nasmyth's theory) was *now* in such a state. Furthermore, it is presently approaching a new equilibrium as its atmosphere cools and attempts to precipitate onto the still-hot 'surface', only to revaporize when it makes contact. Nasmyth blamed the sometimes chaotic appearance of the jovian belts on this sort of boiling turmoil. In other words, the very presence of clouds pointed toward a hot 'surface' condition. White and dark spots were the product of surface volcanoes in the scenario, 'vomit[ing] forth volcanic ash clouds' that reach an altitude visible to us[15]. While Nasmyth could neatly fit observed jovian features into this sort of terrestrial cosmogony, his model for Jupiter required a physical environment quite unlike what we now know to be true.

Nasmyth's conclusions can be justified by the fact that he generalized all bodies in the solar system into a single archetypical planetary model. Nasmyth could not be expected to realize that it might be possible to develop a planetary taxonomy and that Jupiter represents a new physical class of planets, the 'gas giants', which inhabit the outer solar system and differ markedly from the inner, smaller, terrestrial planets (of which the Earth, the planet that Nasmyth was, of course, most familiar with, is a prime example) in meteorological conditions, composition, and evolution. With this *caveat* in mind, Nasmyth's Jupiter-as-ancient-Earth view can be appreciated as both bold and clever. It was also physically plausible, for the entirely optical data base available more than a century ago.

Note that Nasmyth's heat source for Jupiter's atmosphere is endogenic (originating from within): it is primordial and originates entirely from inside the planet. To Nasmyth and others, Jupiter's remoteness from the Sun, and lack of any (observed) seasonal variation, meant that the Sun was an insignificant source of energy for that world. Another way to say this is that, at Jupiter's distance, the Sun could not drive the modest atmospheric changes evident on the contemporary *Earth*—much less the (perceived) enormous and hurried activity on the giant planet.

(Because of Jupiter's size, rapid albedo turnabouts there‡ implied to awed observers the presence of huge energy resources—in order to move great masses great distances. That the quick changes they saw might be illusory, or

---

† For a given density, mass is proportional to volume. The surface area of a sphere increases as a function of its radius squared; the volume of a sphere increases as a function of its radius *cubed*.

‡ Maria Mitchell thought she saw significant, intrinsic changes on Jupiter within half an hour! (Mitchell M 1873 Notes of Observations on Jupiter and its Satellites *American Journal of Science and the Arts* **1** 38.)

Figure 9.2: Richard Proctor. (Courtesy of the Royal Astronomical Society.)

due to subtle phase transitions in a very *thin* layer of the jovian atmosphere, was not adequately considered.)

The idea that Jupiter still is very hot was from Nasmyth, but its 'bulldog' was Richard Proctor (1837–1888). After matriculating at Cambridge, Proctor had switched from banking, to theology, to writing; none were successful careers. Rather, Proctor found a home in speculative astronomy. While few of his ideas remain in vogue today, he is remembered for his books. They not only promulgated his views but also synthesized astronomical knowledge in a readable manner.

Proctor first wrote about Jupiter in an 1870 *Nature* article[16]. He defended Nasmyth's hypothesis, against counterarguments that the energy source driving Jupiter's observed surface dynamics is exogenic (originating from without— like the modern Earth), in this way: if solar energy disturbed the equatorial zone, it could be expected to do so in much the same manner in which the Sun's warmth produces terrestrial storms. These storms build up throughout the Earth day due to the accumulation of radiant heat. Thus, it may be clear in the morning and cloudy in the afternoon. On Jupiter, this process should manifest itself as breaks in the band of clouds encircling the planet, breaks that reflect a daily cycle. However, the turbulent jovian clouds seem often to be continuous—only changing in appearance near the limb due to foreshortening. Therefore, jovian cloud build-up apparently is not subject to a diurnal, but rather a continuous process, such as the release of internal heat reserves.

Thomas Webb, who monitored the reddening of the equatorial zone at the beginning of the 1870s, continued to defend the exogenic origin of Jupiter's heat. He concurred that the zones are the product of atmospheric condensation and form bands because of the velocity of rotation of the planet, but he thought the EZ to consist of two layers of clouds, one reflecting 'yellow' and another white[17]. The contemporary colourization could then be attributed to a thickening of the yellow atmospheric layer by the Sun's radiation.

Yet would not this model require that the equatorial zone be of the deepest yellow colour at the east and west limbs? Even Webb agreed that this did not seem the case[18].

(Johann Mädler and Wilhelm Beer had much earlier stated that the *belts* were subject to acute limb darkening, because spots in them could not be seen beyond approximately 54° from the central meridian. Webb agreed that spots so located are harder to see—markings are foreshortened in the far west or east, while a continuous belt may not be—but saw no appreciable darkening in the belts.)

Webb believed the relative 'transparency' of the polar regions to be evidence that zonal thickness correlates with insolation, that is, cloud formation by exogenic energy. If zonelike bands form in a polar region to the extent that they do in the more temperate and tropical latitudes, their thicknesses (which Webb presumed to be great) produce a perspective effect in which, as we view these latitudes more obliquely toward the poles, the belts (openings, in Webb's theory) between the zones should be increasingly blocked from view. This would imply a more opaque (and brighter) cloud cover at the poles than at the middle latitudes. This is not what is observed. If the duller SPR and NPR are instead devoid of clouds, this suggested that they are very sensitive to insolation or that there is an absence of vapour there to condense into clouds[19].

There were other problems with a Sun-heated Jupiter. On such a planet, others argued that the climate still must be decidedly un-Earth-like. Jupiter's uniquely small obliquity yields little change in the incident solar flux at a given latitude during the course of a year. Its seasonal temperature change was calculated to be proportional only to that experienced during a fortnight in March or September on the Earth. (In contrast, the only major alterations observed on the surface of Mars were seasonal effects.)

By the time American spectroscopist Henry Draper (1837–1882) published his paper entitled 'On a photograph of Jupiter's spectrum showing evidence of intrinsic light from that planet'[20] in 1880, an internally heated Jupiter was the theory to beat.

9.3   Whither the Jovians?

Both Johannes Kepler and Giovanni Cassini speculated that Jupiter might be the abode of life. In 1698, Huygens presented a giant planet with moisture-laden clouds and enough water so that the darker regions on the planet (the belts) could certainly be vegetation—possibly with animals feeding on it. (The clouds hovered over the jovian 'seas' as the white zones[21].) He also spent much time discussing the characteristics of the sentient population of Jupiter.

Subsequent observers of Jupiter speculated on its nature as a home world, too. William Smyth considered the planet a hospitable place of 'eternal spring and serenity, undisturbed by summer's heat or winter's cold'[22]. Science philosopher William Whewell (1794–1866) attempted to justify Jupiter's low mean density by calling it a vast sphere of ice and water, possibly with a nuclear core of low-density cinder. This liquid globe might be the home of 'huge gelatinous monsters languidly floating in icy seas'[23]. Both ideas were considered fanciful by the 1870s but reflected a popular desire for finding habitats suitable for life elsewhere in the solar system. This wish, in turn, was part of a metaphysical approach to planetary science that looked for zoocentric purpose in the Universe. By logical argument from the scarce data, to conditions capable of supporting known life forms on the Earth, many widely differing models for Jupiter could be sustained†. Cautioned one student of Jupiter,

> When we find a theorist gravely arguing from one class of analogies that Jupiter is inhabited by giants fourteen or fifteen feet high‡, while another shows, with at least equal force from other premises, that his people must be pigmies of thirty inches [because of the presumed high surface gravity]; we see at once how futile, not to say absurd, such theorizing is, and how vain is the idea that the purposes of Creation are limited to such objects as we can understand[24].

The source of the above sentiment was the Reverend Edward Firmstone (1823–1900), an Oxford-trained headmaster from Winchester. Firmstone could imagine a Jupiter far from Earth-like and therefore was able to free himself from some of the constraints self-imposed by others.

Firmstone's intent was to put forth a model of Jupiter based upon current knowledge of the planet[25]. He began by delineating two distinct classes of

---

† Professor John Tyndall (1820–1893), of the Royal Institution, was one of the few who still thought that Jupiter might be very cold, owing to its distance from the Sun. But even he admitted the possibility of certain atmospheric gases initiating 'greenhouse' warming. 'Such an atmosphere would store up the sun's heat, and might make the distant planets inhabitable'. (Sharpless I and Philips G 1882 *Astronomy for Schools and General Readers* 3rd edn (Philadelphia: Lippincott).)

‡ The 'argument' goes like this. Pupil size is proportional to eyeball size. Eyeball size is proportional to body size. (Compare cows to people.) Because Jupiter is so much farther from the Sun than the Earth, it is much darker there. Hence, inhabitants need larger pupils in order to see and, therefore, must have big eyes and proportionately taller bodies.... (Ledger E 1882 *The Sun: Its Planets and their Satellites* (London: Stanford)

planets in the solar system: there were those large planets like Jupiter with swift rotation rates, long sidereal years, numerous satellites, and densities equal to or lower than that of the Sun. There also were those tinier bodies like the Earth with longer days, shorter periods of revolution, high densities and few, if any, satellites.

Pluto had not yet been discovered, so the solar system of the mid-19th century was divided neatly, by the asteroid belt, into an interior and exterior group of four worlds each. As these groups differed not only in number of satellites, but also in most other observable respects, Firmstone (and others) were comfortable in extrapolating differences in unobservable characteristics. Because the Earth and Jupiter lay in different groups, and thereby different classes, extrapolation from the Earth was no longer required.

Firmstone (echoing Proctor) avowed that 'surface' conditions precluded the placement of life on Jupiter:

> And why should this shock or surprise us when we remember that there was a time when, in the words of the Holy Writ, 'the earth was without form and void'?—a time which the researches of geology convince us must be measured by millions of years, during which the earth was going through unnumbered changes of temperature— slowly cooling down as some maintain to its present state, in which it is probable that only an outside crust, comparable to the shell of an egg, still hides from us the seat of subjacent fires.
>
> Is it not possible that Jupiter is still in the state from which our globe has emerged, and that the Spirit of God has not yet there brooded over the face of the waters and evoked order out of chaos?
>
> If it be asked—what purpose then does he subserve in creation? the obvious answer is—we are not bound to find a utility cognizable by our finite intelligence for all the Infinite Creator's works. Even of the forces of nature with which we are best acquainted, do we not see by far the larger portion apparently wasted? Rain falling in the sea or on the desert, heat and light visiting the barren mountain top as well as the fertile plain, and radiating continually into space where only a few casual rays are intercepted by the planets? Does it follow that all is wasted whose application we fail to trace; or that the Creator's work is not all very good, because we cannot see it so, as He does?[26].

But what of the future? What if Jupiter were to become more Earth-like? Even in the 19th century, imaginative people considered the possibility of populating other planets with Earth emigrants.

Edmund Ledger noted that Jupiter's oblate shape would affect the strength of local surface gravity. Also, its size and rapid rotation would result in a significant centrifugal force at the jovian equator. This led him to muse about the mental health of transplanted jovians:

If, therefore, the planet should at any future time be inhabited, although there would during the year be very little change of season in any given latitude, it would not only be possible to obtain a change of climate by travelling north or south from any particular locality; but, if any unfortunate individual should suffer from a depression of spirits, he would not only get warmer, but very appreciably lighter by approaching the equator. If, on the contrary, a physician should find his patient to be too elated, he might bid him travel towards one of the poles, with the reasonable hope that his increased weight might steady his feelings. And if temperance principles should not obtain upon the planet, it might be a good rule only to dine out with friends living nearer to the equator than one's own residence, so as to have the advantage of pole-ward movement, and a consequent augmentation of weight and stability, in returning home after dinner[27].

9.4   Earth-like Or Sun-like Joviophysics?

If Jupiter was not (now) necessarily Earth-like, what other body did it resemble? In size and mass, its only master in the solar system was the Sun itself. Firmstone commented that the mean density of Jupiter was only one and one-third that of liquid water (a mere quarter of the average density of the Earth) but was nearly the same as that of the Sun[28]. (Firmstone warned that a weak link in this analogy was the determination of Jupiter's mean density— its radius being uncertain to within perhaps thousands of kilometres because of the unknown potential thickness of an atmosphere.)

Cassini had likened the speedy formation and dissolution of jovian spots to that of sunspots[29]. An atmosphere—solar or jovian—may alter its appearance more swiftly than land or oceans.

Samuel Schwabe also equated jovian spots with sunspots[30]; he even claimed to see **penumbrae** surrounding spots on Jupiter, just as they do on the Sun[31]. Johann Zöllner (ahead of his time; see below) acknowledged Jupiter's differential rotation— particularly near the equator—and noted this similarity with the Sun[32].

George Bond pointed out that Jupiter is brightest at the centre of its disc (not at the limbs) like the Sun. Bond, a fan of an 'Earth-like' Jupiter, wondered whether there were bright aurorae on the giant planet[33].

Another attribute that Jupiter shares with the Sun (as opposed to the terrestrial planets) is its considerable atmosphere. The extent of this atmosphere was clear to observers who witnessed the slow and sometimes oscillating diminution of starlight as Jupiter occulted a star[34]. Only an appreciable atmosphere could account for such gradual extinction. In contrast, when an airless world such as the Moon covers a star, the star's disappearance is abrupt. A Galilean satellite also will fade out as it begins to pass behind the jovian limb. Such reports date from as early as 1823. Charles Todd claimed to spot satellites *through* the edge of the atmosphere at the limb[35]. Similar phenomena had been glimpsed as early as 1857[36]. Furthermore, other observers commented that shadows crossing belts and zones occasionally became elongated, irregular, magnified or disappeared entirely—all indicators of varying depths to the reflecting surface[37]. This all intimated an atmosphere complex, dense and thick enough to have significant inhomogeneities.

For instance, one observer claimed to see white cumulus-like spots *in relief* floating over the 'surface'[38]. That there were clouds on Jupiter had long ceased to be controversial: white clouds helped explain Jupiter's high albedo. Without shiny clouds, how could Jupiter be so bright? (Mars, so much closer to the Earth, should be brighter than Jupiter—all else being equal.)

The brightness of Jupiter had even greater ramifications. In 1865, photometry pioneer Johann Zöllner estimated Jupiter's albedo to be inordinately high (0.62)—even for a planet partly covered by white clouds[39]. George Bond (in 1860) had put it still higher: greater than 1.00[40]! Even Zöllner's more reasonable figure was an average. The zones must be *incredibly* bright to compensate for the belts†.

There were other indications that Jupiter was excessively bright, too. Satellites of Jupiter seemed to have reasonably high albedos themselves; they are very bright against the black background of space. However, when in transit, they seemed to become dark, disappearing for a time near the limb as they became undiscernable from the planet, and then reappearing as *dark* spots as they crossed the disc[41]. (Cassini, his nephew, and the Bonds had all seen such a thing[42].) This is an effect of contrast on the eye. It indicated that the seemingly bright satellites are really much darker than the extremely bright planet.

A mechanism other than reflectivity alone might be needed to account for Jupiter's lustre. Jupiter might have an intrinsic light source. Was the planet partially *self-luminous*? (Accordingly, Jupiter might illuminate its own satellites!)

This radical departure from any Earthly phenomenon was abetted by open comparisons made between Jupiter and the Sun. The theory saw Jupiter as a miniature Sun, though one not nearly as massive or as bright as Sol. (It, too, was encircled by a family of 'planets'—the jovian satellites.)

Cassini had said, 'And we can now doubt . . . that *Jupiter* or his *Attendants* have any other Light, than that, which they receive from the Sun'[43]. In fact, the idea of a luminous Jupiter had received little support since the time of Anaximander (610 BCE–*circa* 546 BCE)! What observational support was there now for the thesis?

First, there were the shadows produced during satellite transits. These shadows were not always completely black[44]; every so often they were 'reddish-

† Regrettably, photometry goes beyond the scope of this book. Bond made his estimation based on his own photographic work, and that of Warren De La Rue. (Bond was the first to make a daguerrotype of any of the planets: Jupiter, in 1851.)

brown or chocolate coloured'[45]. These are the colours one would associate with a dull, red, *glowing* 'surface' of Jupiter. They implied 'a heat similar to that of a red hot iron, perhaps—as its small specific gravity would indicate— still mainly, if not wholly, fluid, still seething and simmering with primeval heat, and even sending forth those vast masses of vapour which cause the appearances visible to us ... '[46]. Jupiter, still cooler than the Sun, might not emit a *lot* of light, but instead radiate a great deal of *heat*†. Thus, conditions on Jupiter could be concluded by observational methods, very similar to those envisioned by Nasmyth through evolutionary arguments.

The question of a suitable energy source to explain observed phenomena was only moved back one step by supposing a hot Jupiter. What now was the source of the energy necessary to heat *the planet itself*? Nasmyth used the bulk properties of Jupiter and his theory of the life-cycle of the planets (contraction?) to resolve this. Firmstone avoided the question altogether:

> Whence this heat comes is a question ... beyond our present knowl-
> edge to solve. When we have ascertained the cause of the sun's heat,
> and *its* unfailing continuance, it will be time to speculate as to that
> of the great planets[47]‡.

Returning to the realm of planetary modelling, John Browning, already known to us as a telescope maker and as an observer, also turned to theory. He further explored the analogy of Jupiter to the Sun[48]. He claimed that the Sun has an 'equatorial belt', 16° wide, in which no sunspots materialize. Such spots seemed, to Browning, to be confined to 'belts', each twelve degrees in width, either side of the solar equator. This situation reminded him of the equatorial zone, with equatorial belts to the north and south of it, in the jovian mid-latitudes. Now, sunspots are comparatively dark regions on the Sun. The 'belts' in which they appear, then, could be considered to have an integrated brightness less than that of the spot-free 'equatorial zone'. Thus, Browning was able to show how the Sun could be equated to Jupiter in terms of its spatial variation in brightness.

(Celestial mapmaker Jean Chacornac[49] (1823–1873) had earlier remarked that Jupiter's spots share another characteristic in common with sunspots: he felt that jovian spots tend to arise in groups just as sunspots do[50].)

Browning did not make the above link on theoretical grounds alone. He already had suggested that the equatorial zone darkening of 1870, which he

---

† In 1874, Zöllner's student Hermann Vogel (1841–1907) would show the jovian spectrum to be nearly identical to that of the Sun. He thereby proved, on astrophysical grounds, that Jupiter shines primarily by reflected sunlight. However, Vogel conceded that an unfamiliar dark band at the red end of the spectrum might be produced by a Jupiter with a consider- able effective temperature. (Vogel H 1876 Untersuchungen über die Spectra der Planeten *Annalen der Physik und Chemie* **158** 461; Hermann D 1973 *The History of Astronomy from Herschel to Hertzsprung* (Cambridge: Cambridge University Press).)

‡ Camille Flammarion, trying to 'save the phenomenon', proposed that Jupiter might emit light visible from the Earth through auroral activity. He did concede that these would be very strange auroras indeed, seemingly affecting equatorial latitudes more than the poles. (Flammarion C 1872 Variabilité de l'eclate de Jupiter *Les Mondes* **28** 555)

made famous, was a manifestation of a cyclical series of such disturbances with a period of about 11 years[51]. Eleven years is also the mean sunspot cycle period. Were these two cycles somehow physically connected? (See below.)

Browning's explanation first assumed that the new colours of the equatorial zone were generated mostly below a neutral cloud layer[52]. He considered these white, mostly opaque, 'cumulus' clouds to maintain a more or less permanent presence in the EZ but that they could be periodically 'left behind' (that is, cleared from certain longitudes) by the rapid rotation of the planet. Still, he could not connect this with distant solar phenomena.

The equatorial zone activity of 1872 caused Proctor to examine the condition of Jupiter's current atmosphere. He began observing Jupiter himself with a Browning telescope, made available to him by a patron†, and concluded that the atmosphere must be quite enormous[53]. An atmosphere the thickness of the Earth's would not be detectable with any instrument at the orbital radius of Jupiter. Yet, Jupiter's was clearly discernible to even the casual observer. Consider, instead, an atmosphere *more than 100 miles deep*, with clouds floating over it, said Proctor. (The cloud tops would be the limit to the observable extent for any atmosphere, which may rise in a transparent state even higher.) Such an atmosphere would be Earth-like, not in direct measure of its height, but instead in proportional thickness to the size of the planet.

Proctor investigated the jovian atmosphere by first creating a 'straw man' model based upon three assumptions: (i) it is of the same composition as that of the Earth; (ii) clouds form on Jupiter at the same pressure as they do on the Earth and (iii) the jovian atmosphere is at least 100 miles deep beneath the clouds. Thus, each assumption rested upon an Earth analogy. Calculations based on Newton and the Ideal Gas Law showed that such an atmosphere would quickly amass a pressure of 10^{20} ('the proportion contains twenty-one figures', in Proctor's nomenclature) atmospheres at the planet's 'surface' (100 mile deep level)[54]. Proctor found this staggering number untenable because at some then yet unknown pressure limit far below this figure, 'air' would be turned into a liquid or solid, a situation he wished to avoid.

Proctor would not allow himself to consider extreme phase transitions within Jupiter and therefore argued that his initial assumptions, which yielded this illogical conclusion, must be false. By doing so, he was able to arrive back at his original tenet that Jupiter is far from Earth-like. Proctor acknowledged that his third assumption could be modified: Jupiter's atmosphere might, in theory, be somewhat thinner than one hundred miles. Still, he objected to reducing it by the order of magnitude necessary to generate a reasonable 'surface' pressure[55].

The astronomy popularizer still was convinced that there was a 'surface' of Jupiter, in the traditional sense, under the clouds. He objected to a homo-

† Lord Lindsay.

geneous Jupiter and believed that its mean density was just that—an average. He discounted old books that spoke of a more or less uniform planet having the density of oak or other woods[56].

Proctor took more time dispelling the imaginative idea, for explaining the giant planet's anomalously low mean density, that Jupiter is hollow! A shell could be made of common terrestrial components and still produce the required low density if it were thin enough. But he pointed out that such a shell would have to be far too thin to withstand gravitational collapse. In fact, with Jupiter's mass accelerating it, such a shell would flow like water[57].

To avoid phase transitions, Proctor introduced the proposition of high *temperature*. In other words, Proctor used Jupiter's atmosphere to substantiate further his hot evolutionary model and the 'miniature sun' theory[58]. Indeed, Proctor's hot Jupiter might be *very* hot. Its colour temperature may be 'cooled' by intervening clouds and gases in the atmosphere. Jupiter's intrinsic energy source may be so mighty that the planet can maintain itself against gravity, solely by its own heat. 'Jupiter is a planet altogether unlike our Earth in condition, and certainly unfit to be the abode of living creatures', said Proctor[59].

The mystery was 'solved'. Proctor took Todd's reported 1877 observation of a satellite through Jupiter's atmosphere as proof of that atmosphere's thickness. The equatorial zone must have turned red because of the thinning of ubiquitous jovian clouds, and the partial exposure of the glowing 'surface' far, far below.

'Not so!', thought selenographer Edmund Neisen (a.k.a. Nevill†; 1849–1940). A literal interpretation of Todd gave Neisen a depth for the jovian atmosphere equal to more than '1000 miles' below the visible clouds[60]. While he was willing to consider an atmosphere made of elements lighter than the Earth's, *or* incredible temperatures at the jovian 'surface', Neisen could not accept *both* these deviations from a standard Earth model.

To Neisen, observations overruled Proctor's theory: Jupiter had white clouds, therefore its atmosphere was made of air and they of water. The bottom of the atmosphere could not be extremely hot; otherwise, the clouds would vapourize. (He used an atmospheric temperature gradient similar to the Earth's; the jovian clouds were set at 0° Celsius—the freezing point of H_2O.) Neisen summarized his displeasure with Proctor's Jupiter by saying, '... it is obviously out of the question to suppose the jovian atmosphere to consist of incandescent hydrogen'[61].

Neisen accused Proctor of requiring a Jupiter so hot that it would turn night to day[62]. Proctor dismissed Neisen's empirical calculations as inappropriate for Jupiter[63].

Yet the Todd observation was inherently impossible, responded Neisen[64]. Io and Ganymede's discs should have appeared flattened, and their perceived

† 'To avoid the stigma of a science career for a member of the old nobility' (Cocks E and Cocks J 1995 *Who's Who on the Moon: A Biographical Dictionary of Lunar Nomenclature* (Greensboro, NC: Tudor).)

occultation times delayed, due to jovian atmospheric refraction. Besides, atmospheric extinction at Jupiter should drastically have reduced the brightnesses of the satellites. Todd had reported none of these phenomena.

Neisen conceded, regarding the second point, that the clouds might not be homogeneously distributed throughout the atmosphere[65]. They might be at the top only. In that case, the cloud layer must be very tenuous so as not to blot out the satellite temporarily, and cause a double occultation event. Yet this layer cannot be *too* thin—else Todd would not have seen it framing a satellite on one side, while the bulk of the planet did so on the other! Such clouds were unlike any *Neisen* had ever seen. (Interestingly, neither Neisen nor Proctor questioned the verisimilitude of Todd's observation.)

Proctor took Neisen's comments as ridicule and suggested that, instead of using constant citations and repetition to back his thesis, Neisen try doing some of the mathematics himself[66]. From here the men's exchange of letters in the *Astronomical Register*[67] turned ugly ...

As is true in many arguments, both participants were wrong. The initial condition of a jovian *surface* brought both Proctor and Neisen to a paradox. A cooler jovian interior required a small solid nucleus of unacceptable density; a hotter interior required a larger, less-dense nucleus of unacceptable luminosity. Yet had either known that Jupiter's atmosphere is made almost exclusively of hydrogen and helium, and accepted the failure of their equation of state when that atmosphere turned into an incompressible fluid (under fantastic pressure), Proctor or Neisen might well have produced an elementary model of the jovian atmosphere similar to what might be generated today.

Parallels drawn between Jupiter and the Sun were easier during the time of Proctor and Neisen because the exploration of solar models going beyond perfect gas behaviour did not occur until the late 1870s. As recently as 1907, both Jupiter and the Sun still were thought to possess some kind of hot 'surface' or crust, albeit far below an extensive fluid envelope[68]. Jovian volcanoes still were mentioned enthusiastically in the literature late in the 19th century[69]. The speculated distance from the top of the atmosphere to the unseen 'bottom' continued to increase, however.

Jupiter was allowed another step toward revealing its gaseous, nonterrestrial nature by John Brett, who measured the thickness of the planet's atmosphere in thousands rather than hundreds of miles. Brett believed that the large white spots he described in 1876 were most likely globular in shape with only their projected two-dimensional discs visible to us on the Earth. One of these discs had a diameter of '6000 miles'[70]. Assuming a spheroidal shape, this meant that the supposed shadows projected by these globules could be on a 'surface' at least 6000 miles below its top. (Brett considered the features to be totally immersed in the atmosphere because they were never seen rising above the planet at the limb.) Between the top of the globule and the planetary 'surface' then must be an atmosphere thick enough to float the globule.

Brett's idea of 6000 mile across spheres 'swimming' through the jovian atmosphere was an impressive one. He was willing to go even further. '...if Jupiter *has any nucleus at all* [my italics]', said Brett, 'it is not visible to us'[71].

A textbook published in 1882 gives a summary of the contemporary jovian model:

> There is strong evidence that Jupiter is still very hot. The sun's rays are too feeble there to raise the thick clouds which constantly envelop him, and the great changes continually going on in his atmosphere must be caused by intense heat within. The earth shows plainly that it was once in a molten state, and if it was created at the same time as Jupiter, the great mass of the latter would keep him hot long after the earth had cooled off. If the body of the planet has yet solidified, it is still probably white-hot. So that Jupiter is more like the sun than like the earth. And indeed there is evidence that he actually gives out some *light* of his own, even through his dense cloudy atmosphere. The amount of this, however, if any, cannot be great: the most of his light is reflected sunlight[72].

9.5   The Search for Periodicities in the Jovian Atmosphere

The cyclical nature of things in the heavens is one of astronomy's most obvious truths: planets revolve through the zodiac; the Moon waxes and wanes; the seasons come and go and come again. This fact is so engraved into the science that it is normal to search for further natural periodicities. The dynamics of the jovian atmosphere are not an exception.

Yet the mechanism for seasonal effects on the Earth, the angle between the planet's planes of rotation and revolution, had already been dismissed as ineffectual in the case of Jupiter. What other means for regular change was there?

Browning's suggestion[73] that disturbances on Jupiter were periodic did not go unnoticed. John Herschel had been the first to propose that such periodicity might be attributable to the sunspot cycle. Browning now came forward with a sketch he had made in January 1860, with star-cataloger Charles Pritchard's[74] (1808–1893) $6\frac{3}{4}$ inch Cooke refractor. Browning's notes for the sketch indicate that it was precipitated by peculiar markings and the 'remarkable brown colour' of the equatorial zone. The southern edge was 'divided into egg-shaped masses, which must have been of brighter or lighter colour than the background of the belt'[75]. Browning equated this description with that of one of his drawings made in January 1871. He placed a request in the *Monthly Notices of the Royal Astronomical Society* for observations of colour in the EZ prior to 1860.

Arthur Ranyard (1845–1894) was a Cambridge-educated barrister intrigued

Figure 9.3: Arthur Ranyard. (Courtesy of the Royal Astronomical Society.)

by the regularity of eclipses and their observation. His most notable astronomical accomplishment was to come later when he compiled a comprehensive volume of eclipse data for the RAS. His interest in historical data bases is evident in the tack he took with supposedly new jovian phenomena such as Mayer's ellipse and Browning's equatorial zone reddening.

Looking at drawings made by William Huggins, William Lassell, George Airy, and Bruce Murray during the period 1858–1860, Ranyard identified what he contended to be an overabundance of features compared to other times. He interpreted this as evidence for the repetition of a cycle of jovian activity and quiescence with the same period as the sunspot cycle† (11 years). Searching through the index to the *Monthly Notices*, recently prepared, Ranyard compared information on Jupiter to Rudolph Wolf's (1816–1893) Zurich sunspot count. He found the initial reports of Lassell's spots in 1850; the sunspot maximum was in 1848. There were no *Monthly Notices* during the previous sunspot maximum in 1836–1837. The three sunspot maxima earlier than this were 'inferior', and there were no reports of spots on Jupiter between 1795 and 1830[76]. However, Ranyard did locate some notes by William Herschel

† Samuel Schwabe was the first to demonstrate that there exists a periodicity in the average number of sunspots visible on the solar photosphere at a given time.

describing white spots in the jovian bands during 1778–1780. This period
was again concurrent with the sunspot maximum. This particular maximum
was a lengthy one and enabled Ranyard to fit Schröter's 1786 observation of
'remarkable changes' on Jupiter into the correlation[77]. Likewise, the sunspot
activity of around 1693 was coincident with Cassini's 1692 published observation of jovian features.

Was there really a relationship between Jupiter's atmosphere and sunspots?
Astronomers had looked for such a 'solar-jovian connection'. The near com-
mensurability of the sunspot cycle and the jovian year may have prompted
the inquiry initially. Unfortunately, the jovian year is also the period of optimum jovian observing conditions for a given Earth hemisphere. (This causes
a modulation of resolution inescapable for observers unwilling to adopt a mi-
gration pattern of 12 years.) Furthermore, a mechanism for interaction that
would span five astronomical units was unknown. Later, there would be those
who were tempted to invoke magnetic phenomena or the solar wind to bolster
what seems a coherent argument so far, but such effluvia were not well understood in the 19th century. Then, tidal forces seemed the only reasonable
explanation put forth, though Ranyard himself disapproved of this idea[78]. He
sensed the handiwork not of one body on the other but of some independent
agent acting simultaneously on both[79]

The polemical counterargument that Ranyard met was much the same as
that encountered by Browning when he had said that the equatorial zone fundamentally changed its appearance. RAS members suggested that if Jupiter
had always been studied with similar instrumentation, it would always have
appeared much the same. This criticism caused Ranyard to make more rigorous his argument: 'In answer to such objections I propose to compare a small
series of drawings, all made within the last twenty years, by well-known observers with large instruments'[80]. To begin, he produced Lassell's 'very poor
woodcut' of 1850, which nonetheless demonstrated that the white spots and
'broken condition of the equatorial region, are very apparent'[81].

'Within a year of the next Sun-spot [*sic*] minimum Mr W De La Rue made
his large and well-known drawing of *Jupiter*', which according to Ranyard was
'universally acknowledged to be a very just representation of the state of that
planet'[82]. In it, there are no signs of 'Dawes markings'[83]. Neither do they
show up in Piazzi Smyth's Tenerife drawings. Ranyard put particular trust
in Smyth's work because he made his observations when Jupiter was nearly
at the **zenith**.

Sunspot maximum occurred in 1859. In November 1858, Lassell rediscovered his spots. Ranyard did not agree with William Dawes' assessment that
they had been in continuous existence. Ranyard believed that Lassell's spots
behaved more like sunspots, which do not last from one cycle to the next. He
used this as another example of simultaneous jovian and solar activity[84].

Ranyard proceeded to the 1860s and along the way described an 1859 picture of Jupiter 'beautifully drawn by Sir K Murray at his 9-inch refractor'[85].
In it he saw 'flocculent and cloudy port-holes [*sic*] in the principal belts'[86].

(These spots were apparently not seen by *Astronomical Register* founder San-ford Gorton (1823–1879), who watched Jupiter through a modest $3\frac{1}{2}$ inch Cooke simultaneously.) Ranyard went on to reference the Astronomer Royal's report to the Board of Visitors for 1860–1861. In the same drawing by Car-penter that Airy had used to illustrate his point about the constancy of colour on Jupiter, Ranyard found it 'showing all the flocculent port-holes [*sic*] and reddish colour of the equatorial region, bright eggs, and elliptic markings, which have been so noticeable during the last two years'[87]. (!)

The most recent sunspot minimum was in 1866. Unfortunately, Ranyard discovered that few drawings of Jupiter had been made during or near this year. Ranyard used a new argument here: he stated that this scarcity might be confirmation in itself of the lack of significant markings to attract astronomers' attention[88].

If they did not draw, perhaps they wrote something in their observing logs. Webb was known to have carefully observed Jupiter during the mid-1860s. He also was interested in jovian atmospheric morphology. Ranyard asked him to review his notes[89]. There, only three spots were recorded between June 1863, and August 1866. None of these was the big equatorial oval of 1870. It is not clear whether Ranyard requested records of spots or of features in general.

Ranyard's *coup de grace* was that during the recent jovian activity, sunspot counts seemed to be peaking[90].

In 1874 Lassell once again recovered his white spots. He lent his authori-tative voice in support of periodic phenomena, 'which', he said, 'I think more probable than that absolute secular changes occur in the heavenly bodies within the limit of time of any human record'[91].

While others were looking at what was going on with Jupiter at sunspot maximum, George Hirst investigated events at sunspot minimum. He pointed out that Warren De La Rue's and Piazzi Smyth's drawings, made at the minimum, are devoid of white spots. In 1876, the year Hirst himself began observing Jupiter systematically and also a sunspot-minimum year, he found white spots rare (though others saw some)[92].

If white spots develop at sunspot minimum, what did the tiny dark spots that could be seen in Jupiter's belts from time to time foretell? Hirst felt that these features (minute and observable only with a sizeable aperture) might coincide with sunspot minimum. It was still proposed by some that these black spots were actually jovian surface features—mountaintops that occasionally managed to poke above the jovian clouds. Hirst dismissed this idea by saying that their albedo was much lower than what would be expected for any geological feature[93].

Next, Hirst went back to Browning's original suggestion that jovian colour be examined for repeating patterns of intensity. Hirst could find three periods of vivid colouration. The first was an old observation by William Herschel, who designated the equatorial zone yellow in 1794. (This evidently was be-yond the instrumental hue that Herschel's specula may have always given to Jupiter.) The second was the archetypical one of *circa* 1870. The third was

going on as Hirst wrote (in 1877). This most recent reddening exceeded the
abbreviated one reported by Edward Knobel earlier in the decade[94]. (The
journal *Nature*[95] called attention to another historical occasion of 'reddish
brown' in the EZ reconstructed from observations made by Franz Gruithuisen
(Director, Munich Observatory; 1774–1852) between 1836 and 1840.)

Yet, after all his research on the subject, Hirst eventually decided that the
evidence for a solar influence on Jupiter remained inconclusive[96]. He did not
deter others; speculation about a solar–jovian relationship was to continue for
nearly 100 years[97].

Today it seems strange to construct an elaborate documentation of a major
phenomenon (periodicity of jovian features and activity) for which no satisfac-
tory physical explanation had been advanced, as Arthur Ranyard and others
did. This, though, reflected the shift in emphasis from astronomy by observational induction to astrophysics. Contemplation of a mutuality between the
two titans of the solar system, the Sun and Jupiter, was not to be limited to
one astronomer or one set of data. The importance of Ranyard's work is that
it is the first serious use of historical records to study long-term phenomena
on Jupiter. Ranyard was a historical astronomer. This present work is in
many ways a descendant of his.

9.6   Observation Versus Theory

Heretofore I have divided the history of the observations made of Jupiter,
and the history of the theories formulated about that planet, into separate
chapters. Now I would like to bring these subjects together by discussing
the commonality shared by the two endeavours, observation and theory. As
my examples show, the same people often were involved in both, particularly
after the 18th century.

This was not by intent. The scientific style of the amateur planetary astronomer *circa* the 1800s did not call for speculation. It called, modestly, for
an accumulation of the 'facts'. Futhermore, one did not need to understand
much 'theory' to observe Jupiter. The rudiments necessary to write intelli-
gently about features on the planet were made available by word of mouth, or
through popular works put out by the likes of Richard Proctor and Camille
Flammarion. These did not show one where or how to look, either—practice
did.

Moreover, amateur astronomers were not automatically interested in a theory to explain their observations. Frequently, a good observation—a unique
one—was reward by itself. Amateurs often wanted a thing or place to which to
attach their name, not a 'law' or principle. As an example, today, new comets
frequently are discovered by amateurs who scan the skies nightly seeking out
just such an elusive object before someone else finds it. Comets traditionally are named by and for their discoverers. Yet the names of comets and

the citations attached to writings about intrinsic properties of comets seldom correspond.

In their writings, the amateur observers of my study period generally felt it was sufficient to detail their findings. If moved to do so, they might perhaps raise a question or two posed by these observations, either implicitly or explicitly.

I mentioned earlier that there was a dearth of theory that an observationalist needed to know. I can go further: there was a dearth of theory. For if the amateurs were expecting professional astronomers seriously to take their observations in hand and make sense out of them, they were, by and large, disappointed. Professional astronomers of the day were not pursuing theoretical investigations of the planets. This was a time when the early experiments in astrophysics were making stellar physics and celestial mechanics seemingly more fruitful ground for planting their talents. The directions that these new scientific developments would take were clearer. A spectrum can be duplicated in a laboratory; a shading of a cloud on Jupiter cannot.

Jovian theory was something of a 'hot potato'. Even after the fear of the stake was removed, it was not something on which a professional wanted to stake his or her career, considering the nil rewards. Amateurs, though, did not have any professional reputation to conserve at all, and so, theoretical interpretation, too, became, by default, their province. The incidental guesses offered as the epilogues to observational papers slowly were knitted together into something we would call a theory for the planet Jupiter. If it was not clear at its introduction, much of the interpretive work outlined in this chapter was skimmed from appendages to the written work reporting on observing programs described in chapters 3–8. Jovian astronomers were aquiring their notions about the structure, composition, and behaviour of the giant planet almost incidentally from their colleagues. This was possible because the total of certain knowledge remained small during this time. It also was possible because the technical language necessary to present these ideas was, in fact, quite simple.

The Jupiter-watcher of the 19th century accidently became the Jupiter theoretician of that same period, and of the early part of the 20th century. The names were the same. Sometimes, those who tended to proffer the most about the meaning of jovian observations (the Nasmyths and the Firmstones) were individuals who had little or no observations of discovery to their credit. They may well have preferred that the circumstances had been the other way around. Lassell did not have to write treatises interpolating many observations in a search for meaning; he had discovered both planetary features and new satellites.

Proctor's success lay in recognizing the duality of role in the planetary astronomical community. He gathered, edited, expanded upon and made coherent the ideas often hypothesized—however tentatively—by others, and then presented this work back, as resources, to these very same persons.

9.7   Summary

Cool or hot 'surface'? Thin or thick atmosphere? Energy from within or without? The story of the theoretical interpretation of Jupiter is also that of a *yin/yang*-type interaction between two competing paradigms. These paradigms were in the form of prototypes for Jupiter, the Earth and the Sun, objects believed to be better understood than distant and enigmatic Jupiter. Extrapolation from what was known to what was not was a natural and time-honoured thought process.

One paradigm interpreted Jupiter in a way so as to make it as closely as possible a surrogate for our own world. This contrasts with a viewpoint that sees Jupiter as a smaller version of the Sun.

The idea that the jovian atmosphere is a scaled-up analogue of the Earth's predates the scenario that modelled it as a scaled-down solar atmosphere. The latter view only emerged coincident with the birth of solar physics in the second half of the 19th century. The collision between these two ideas was, largely, a *productive* one. Indeed, this competition begun in the last century profitably continues today with the solar analogue somewhat in the lead.

If this approach had not been taken, it is unlikely that a totally independent theory of Jupiter, drawn from nebulous 'first principles', would have come into being at any time before very recently. If Jupiter was not constructed in the image of other bodies—and we are fortunate that this often was done by the same people who contributed the observational data—any attempt at studying its planetary nature certainly would have been retarded significantly.

Endnotes

[All titles are written in full, with the exception of '*Phil. Trans.*' for the *Philosophical Transactions of the Royal Society* and '*Mon. Not.*' for the *Monthly Notices of the Royal Astronomical Society*.]

1. Kuhn T 1957 *The Copernican Revolution* (Cambridge: Harvard University Press) Kuhn's premise is expanded upon in his pivotal *The Structure of Scientific Revolutions* (Chicago: University of Chicago Press)
2. Van Helden A 1974 The Telescope in the Seventeenth Century *Isis* **65** 38
3. Guillemin A, Proctor A and Lockyer J (eds) 1883 *The Heavens: An Illustrated Handbook of Popular Astronomy* (New York: MacMillan)
4. Sheehan W 1992 *Worlds in the Sky: Planetary Discovery from Earliest Times through Voyager and Magellan* (Tucson AZ: University of Arizona Press)
5. Chambers G 1877 *A Handbook of Descriptive Astronomy* 3rd edn (Oxford: Clarendon) Hind was also a product of George Bishop's observatory, as was Albert Marth.
6. Denning W 1883 Observations of Jupiter. *Mon. Not.* **43** 97

7. 1847 Jupiter and His Moons. *The Sidereal Messenger* **1** 73
8. Sheehan W 1996 *The Planet Mars* (Tucson AZ: University of Arizona Press)
9. Huygens C 1698 *The Celestial Worlds Discover'd or, Conjectures Concerning the inhabitants, Plants and Productions of the Worlds in the Planets* (London: Childe)
10. Webb T and Espin T 1893 *Celestial Objects for Common Telescopes* (London: Longmans, Green)
11. Herschel J 1849 *Outlines of Astronomy* (Philadelphia PA: Lea and Blanchard)
12. Terby F 1880 Observations de la tache rouge de Jupiter. *Bulletin de la classe physico-mathematique de l'Academie de Sciences de Saint Petersborg* **41** 210
13. Nasmyth J 1853 Some Remarks on the Probably Present Condition of the Planets Jupiter and Saturn in Reference to Temperature, etc. *The Edinburgh New Philosophical Journal* **59** 341
14. Clerke A 1908 *A Popular History of Astronomy During the Nineteenth Century* (London: Black). Agnes Clerke (1842-1907)
15. *Ibid.*
16. Proctor R 1870 Are Jupiter's cloud-belts due to solar heat? *Nature* **2** 326
17. Webb T 1871 The planet Jupiter *Nature* **3** 430
18. *Ibid.*
19. *Ibid.*
20. Draper H 1880 On a photograph of Jupiter's spectrum showing evidence of intrinsic light from that planet *Mon. Not.* **40** 433
21. Huygens C *Op. Cit.*
22. Firmstone E 1872 The planet Jupiter *Winchester and Hampshire Scientific and Literary Report of Proceedings* 81
23. *Ibid.*
24. *Ibid.* The 'theorist' is likely Huygens.
25. *Ibid.*
26. *Ibid.*
27. Ledger E 1882 *The Sun: Its Planets and their Satellites* (London: Edward Stanford)
28. Firmstone E *Op. Cit.*
29. Cassini G 1672 A relation of the return of the great permanent spot in the planet Jupiter, observed by Signor Cassini, one of the Royal Parisian Academy of Sciences. *Phil. Trans.* **7** 4039
30. Schwabe H 1860 Über die Steifen des Jupiter *Astronomische Nachrichten* **53** 137
31. Webb T and Espin T *Op. Cit.*
32. Zöllner J 1871 Über das Rotationsgesetz der Sonne und der grossen Planeten. *Berichte über die Verhandlungen der Königlich Sächsischen Gesellschaft der Wissenschaften zu Leipzig* **23** 49
33. Clerke A *Op. Cit.*

34. Firmstone E *Op. Cit.*
35. Todd C 1877 Observations of the phenomena of Jupiter's satellites at the Observatory, Adelaide, and Notes on the physical appearance of the planet *Mon. Not.* **37** 284
36. Neisen E 1878 The Atmosphere of the Planet Jupiter *Astronomical Register* **15** 225
37. Proctor R and Ranyard A 1892 *Old and New Astronomy* (London: Longmans, Green)
38. Firmstone E *Op. Cit.*
39. Bryant W 1907 *A History of Astronomy* (London: Methuen)
40. Bond G 1861 On the Light of the Moon and of the Planet Jupiter *Memoirs of the American Academy of Arts and Sciences* **8** 29
41. See e.g., Dancer J 1868 Jupiter as observed at Ardwick on the night of August 21st, 1867 *Memoirs of the Literary and Philosophical Society of Manchester* **7** 10
42. Clerke A *Op. Cit.*
43. Cassini G 1666 A more particular account of those observations about Jupiter, that were mentioned in numb. 8 *Phil. Trans.* **1** 171
44. Hough G 1867 Note on the appearance of Jupiter, August 20th, 1867 *Mon. Not.* **27** 323
45. Firmstone E *Op. Cit.*
46. *Ibid.*
47. *Ibid.*
48. Browning J 1872 The condition of Jupiter *The Student, and Intellectual Observer* **1** 1
49. Bryant W *Op. Cit.*
50. Chacornac J 1865 Taches de Jupiter observèes en Juin par M Chacornac *Les Mondes, Revue Hebdomadaire des Sciences, et de Leurs Applications* **8** 485
51. Browning J *Op. Cit.*
52. *Ibid.*
53. Proctor R 1873 News from Jupiter *Popular Science Review* **12** 348
54. *Ibid.*
55. *Ibid.*
56. *Ibid.*
57. *Ibid.*
58. *Ibid.*
59. *Ibid.*
60. Neisen E *Op. Cit.*
61. *Ibid.*
62. *Ibid.*
63. Proctor R 1879 The atmosphere of Jupiter *Astronomical Register* **16** 19
64. Neisen E 1879 On the atmosphere of the planet Jupiter *Astronomical Register* **16** 42
65. *Ibid.*

66. Proctor R 1879 Jupiter's atmosphere *Astronomical Register* **16** 77
67. Proctor R 1879 Jupiter's atmosphere *Astronomical Register* **16** 130
68. De Vorkin D 1984 Stellar evolution and the origin of the Hertzsprung-Russell Diagram *Astrophysics and Twentieth Century Astronomy to 1950: Part A.* (Cambridge: Cambridge University Press) 90
69. See, e.g. Flammarion C 1877 *Terres du Ciel* (Paris: Libraire Académique Didier)
70. Brett J 1876 On the proper motion of bright spots on Jupiter *Mon. Not.* **36** 355
71. Brett J 1874 Memorandum of observations of Jupiter made during the month of April 1874 *Mon. Not.* **34** 359
72. Sharpless I and Philips G *Op. Cit.*
73. Browning J 1870 Note on the alteration in the colour of the belts of Jupiter *Mon. Not.* **3** 202
74. Herrmann D 1973 *The History of Astronomy from Herschel to Hertzsprung* (Cambridge: Cambridge University Press)
75. Browning J 1871 Note on the change in the colour of the equatorial belt on Jupiter *Mon. Not.* **31** 75
76. Ranyard A 1871 On periodical changes in the physical condition of Jupiter *Mon. Not.* **31** 34
77. *Ibid.*
78. *Ibid.*
79. Chambers G *Op. Cit.*
80. Ranyard A 1871 On the physical changes on Jupiter *Mon. Not.* **31** 34
81. *Ibid.*
82. *Ibid.*
83. *Ibid.*
84. *Ibid.*
85. *Ibid.*
86. *Ibid.*
87. *Ibid.*
88. *Ibid.*
89. *Ibid.*
90. *Ibid.*
91. Lassell W 1874 On the appearance of round bright spots on Jupiter *Mon. Not.* **34** 310
92. Hirst G 1877 Some notes on Jupiter during the opposition of 1876 *Journal and Proceedings of the Royal Society of New South Wales* **10** 83
93. *Ibid.*
94. *Ibid.*
95. 1877 *Nature* **15** 282
96. Hirst G *Op. Cit.*
97. See, e.g. Basu D 1969 Relation between the visibility of Jupiter's Red Spot and solar activity *Nature* **222** 69; Balasubrahmanyan V and Venkatesan D 1970 Solar activity and the Great Red Spot of Jupiter *Astrophysical Letters* **6** 123

Chapter 10

Observations, Conclusions, and Final Thoughts

... in moments of magnificent definition, such a wealth of detail has been pre-sented to my sight that my pencil has lain idly by, and I have been content to gaze in almost open-mouthed wonder.

George D Hirst, Fellow, Royal Society of New South Wales
1876

10.1 A Golden Age

The 19th century, in particular, was a golden age for observing Jupiter and for astronomy in general. The establishment of observatories, some of which were national endeavours patterned competitively after the Paris Observatory while many more were private enterprises, contributed to this.

Similarly, the resources available for, and interest in, publishing single-subject astronomical journals were positive factors. Many periodicals came and went during this era, but those few that held on established themselves with a reputation of being *the* means of astronomical communication. The fact that several prominent survivors were written in English leant credibility to the claim of English as a scientific language and helped English astronomy prosper.

Certainly, though, it is difficult to give too much credit to the construction of the silvered-glass reflector (and the accompanying Focault knife-edge test) in advancing the telescopic examination of Jupiter. This contemporaneous quotation from the day concurs: 'The perfection of the silvered glass speculum has made clear and certain some features which all but the most exquisite achromatics fail to bring out'[1].

Large apertures, produced by John Browning and his like, put high resolution and colour sensitivity at the disposal of a legion of new observers, ironically at a time when the reflector was still disdained by the comparatively few established professional astronomers. Shortly thereafter, planetary (and solar) astronomy, armed with its new instruments, led the exodus away from and above poor observing conditions, to benign climates and mountaintops, for an even clearer view.

Yet perhaps the 1800s are best remembered for events that did not take place at the telescope eyepiece. The improvement of planetary nomenclature and experiments with planetary classification were equally important indoor accomplishments.

Regarding the former, it was the nomenclaturists who caused us to think of Jupiter as a collection of places. New questions were formulated, and old ones were phrased in new ways. What are the latitude and longitude of such a place on the planet? What are its usual albedo and colour? What features might it contain? These individuals also advocated standardization. A goal was to avoid calling a feature one thing when considering it a foreground phenomenon and another thing when considering it part of the background. (Another purpose was to eliminate confusing statements such as: 'The south tropical dark belt is really part of the equatorial zone, but I have given it a distinctive name in consequence of its thinning out, and therefore appearing like a separate belt altogether.'[2]!)

A less-fortunate legacy of this contingent, whose aim was to improve Jupiter language, is the tendency to make systems of nomenclature that describe superficial characteristics too in-flexibly. In other words, their attempts to use a set of band names to describe the belts and zones from one apparition to the next were *too* successful, as the characteristics of these bands, used in the naming systems, do in fact change over such intervals.

The birth of astrophysics did not fail to influence the analysis of Jupiter. Questions of causes and processes occurring there arose and were finally addressed. The study of Jupiter became a reasoned theoretical one as well as one conducted by observation and induction.

Applying the new physics to Jupiter had begun, of course, with universal gravitation's establishment as a governing force in the Universe, but now it was put to work in modelling Jupiter and analysing aspects of the planet that could not be seen. An arsenal of atomism, thermodynamics and kinetics became available to jovian theorists.

Interpretation was the hallmark of the last three decades of the 19th century. Early examples include attempts to place Jupiter in an age sequence with the other planets based on primordial heat retention. They also include the application of atmospheric physics to a thick atmosphere. The inability to neatly fit Jupiter into a model of a small Sun or a large Earth only spurred further investigation. Attempts by astrophysicists to investigate interrelations between the Sun and the Earth led those looking at Jupiter to search for solar–jovian connections, as well. While confusion of cause and effect was to last

100 years, the beginning of systematic investigations of periodic change on Jupiter encouraged the comprehensive observing programmes that continue today.

The questions asked about Jupiter now were first articulated in the 1870s: what are the various depths of observable features? How is the planet organized internally? How is energy received and distributed? Is activity periodic?

These questions have no easy answers. Interest in Jupiter was to subside and be preempted by major advances elsewhere in astronomy during most of the 20th century. However, if Jupiter's history is cyclical—and it seems to be behaving that way—a new era of investigation is imminent as we are poised between the first cursory examinations of the giant planet made close-up and the full-scale visits of exploration that will be made there during the 21st century.

10.2 Who Were the Jupiter Watchers?

A discipline does not proceed autonomously. It is the combined work of numerous individuals.

At first, many planetary astronomers were Italian. The astronomical telescope took root in Italy, and for a long while there was little distinction between instrument-maker and instrument-user. Italian planetary astronomers also were successful because of their knowledge and tradition of soliciting patronage.

Italian planetary astronomy stagnated during a period after the counter-Reformation. After the end of the doldrums of the 18th century, planetary astronomy had sown itself chiefly in the British Isles (by way of Germany).

Among English-speaking astronomers, Jupiter was almost entirely the province of amateurs during the 1800s. Only Astronomer Royal George Airy took an interest in using the facilities of the astronomical centre of England (Greenwich) for a programme of planetary observation of any kind[3].

The amateurs were aided by the technological innovations mentioned earlier: first, the speculum metal reflector, which gave them quality and large aperture at a reasonable price, and then, the silvered-glass reflector, which freed them from the extremely time-consuming maintenance of their speculum.

The number of names hitherto scattered through this book is somewhat misleading. Astronomy was but a temporary diversion for these people, and the cadre at work at any given time was small. The size of their country and its reliable communications network kept them in touch with one another.

What was needed to be an amateur planetary astronomer in the English style has already been touched upon: time, money, and, almost invariably, a liberal arts background (or interest therein in the case of practical men). A deep devotion to objectivity, at least philosophically, was required. A certain amount of mechanical dexterity was also an asset. The reader perhaps has been struck by the many professions represented by these observers. Physicians, lawyers, soldiers and clergy took their turns at the eyepiece. (Few were teachers, the requirement for fiscal resources there being lacking.) In fact, persons with theological training appeared on the scene twice during my study period of jovian observation: persons from the Catholic orders during the first decades of the telescope and then, during the time I am now addressing, Anglican ministers.

While it might be expected that the data taken by this corps of volunteers might be reduced, sorted and theorized about by those in academia, this was not strictly the case. Richard Proctor borrowed upon the expertise of others, but had little formal training himself before attempting theoretical jovian astrophysics. Several prominent observers felt free to at least occasionally speculate about what they saw (though most did not).

Among those who thought about Jupiter's nature as opposed to its appearance, the points of view expressed represent a common metaphysics of finding purpose in Jupiter. The variety of ideas came largely from the variety of opinions concerning for whom Jupiter's 'purpose' was intended. Was it for the benefit of inhabitants of the Earth? Citizens of Jupiter or its satellites? Non-sentient creatures dwelling there? Or was it for the Creator's own plans unmanifest to us?

Above all, the success of (principally) English students of Jupiter in the latter 1800s can be attributed to a single quality: the willingness to cooperate and work together—an unfamiliar idea in the competitive mood of two centuries earlier. For the first time, the observers, first informally and then formally, organized themselves for a common objective, realizing that the subject was beyond the ability of any single astronomer to monitor sufficiently or to deduce its nature alone.

10.3   Describing Jupiter

What was seen, recorded and published concerning Jupiter changed through history as various subjective forces were at work. Today we can recognize a series of—for want of a better word—observational 'fads', rather than a self-consistent catalogue of events, when searching the historical record. (Perhaps the best example is the sudden apparent frequency of Lassell's spots in the mid-19th century.) When certain features were in vogue, we cannot be sure that others were not slighted. This character of the record makes interpretation of it as a single body difficult.

To begin with, any marking on Jupiter that could be made out was unique and therefore a subject of great interest. Relatively few could be seen through telescopes in the early days—and those that were, were described exhaustively.

The sequence in which particular feature morphologies were recognized reflects a descending order of apparent angular size. Thus, the jovian disc itself could be considered the first 'feature' associated with the giant planet. After that, global phenomena, the belts and zones, were made out. Then, smaller and still smaller features were resolved: large spots, curvilinear markings that were longitudinally narrow but extended over a great range of latitudes (initially those in groups and then isolated examples), irregularities in the belts and zones, small spots that came in clusters and finally small individual spots.

The contrast of a feature against its background also affected the order in which phenomena were discovered. Satellite shadows are angularly small but nonetheless were made out quite early. Similarly sized intrinsic spots of less contrast were not documented until 200 years later. Likewise, the dark northern 'barges', which consistently provide the greatest contrast of any regularly recurrent jovian surface feature, were recorded fairly soon.

Short-lived features were among the last identified during my study period. Discussions of band disturbances and the morphologies associated with them began only *circa* 1870. Understandably, the rarer a phenomenon was, the more slowly it came to the attention of observers.

The use for which observations were made influences what we can expect to get out of observing logs from long ago. For instance, at a time when the jovian atmosphere was thought to be quite thin, dark spots were thought to be glimpses of a very real solid 'surface' below it. Yet, interestingly, most often light spots were used by those merely involved in determining the jovian rotation period because they could not be mistaken for satellite shadows, were more plentiful, were easier to pick out and track, and were more likely to be recoverable after transiting the far side of the jovian globe. Indeed, it must be stressed again that, in the nautical age following the 17th century, the jovian system was thought of as of greatest interest as a navigational device, a reason for the decline of more probing jovian studies. This bent can be seen in the rush to produce jovian ephemeri, even before the nature of what was being seen was close to being understood. Morphological observations were often serendipitous.

This leads us to another selection effect. Foreshortening makes features more difficult to see and measure at the limb. Rotation-timers concentrated on the equatorial regions of Jupiter (where markings could be found at least some of the time) away from the edge of the disc, to the exclusion of the polar regions. Of course, this restriction to low latitudes was of no hindrance to the few trying to find similitude between goings-on on Jupiter and terrestrial meteorology and oceanography. These were the latitudes where most global voyages took place on the Earth. Even into the 19th century, when keen morphological interest arose, polar features were missed or not looked for because of the problem of lack of contrast there. News of markings beyond the first temperate bands was welcomed as a novelty. As there was always little to see in the northern hemisphere, attention remained focused on the equatorial zone, south equatorial belt, south tropical zone and south temperate belt and zone.

With quality optics, and large apertures, came the ability to distinguish colour more easily. This, though, does not completely explain the suddenness of the onset of regular colour documentation. Previous references to colour had been scarce and vague, e.g., 'die Aequatorialzone ist hellgelb, wie fast immer' [the equatorial zone is light yellow, as always][4]. Now, late in my study period, accurate colour determination became of utmost importance. We have only come to expect more and more colour and colour commentary since.

This quest for colour fidelity may actually have delayed the introduction of useful coloured filters into the optical path. Nowhere does even the mention of the potential for such filters, to aid in clarifying indistinct low-contrast features, seem to appear. Only 'natural' means for doing this were acceptable: the slowly learned lesson of matching magnification to seeing conditions and techniques for reducing the harshness of background contrast (such as viewing Jupiter during astronomical twilight).

When observers were few and isolated from one another, there was no hope of getting an appropriate sense of the time variability of the planet's appearance. After the late 1860s, enough observations began to overlap so that it did become possible to assemble a truly longitudinal (in its chronological sense) study. Still, to the end of my study period, the question of feature lifetimes was made difficult by the inability to figure out always whether sightings of a 'feature' were each a unique representation of a phenomenon or evidence for the continuation of a single example of that phenomenon. Sometimes this confusion was due to happenings beyond the observers' control (e.g., the declination of Jupiter, solar conjunction and the weather); at other times, a coincidental event on the planet would 'sidetrack' observers from documenting ongoing activity.

For, fundamentally, it was the exception rather than the rule recorded in the case of Jupiter. Only the most unusual things were likely to be included in the journals. The lack of full-time professional jovian observers who, over many years, developed an intimate familiarity with the planet, left the history of observations of Jupiter modulated by a series of 'sudden' discoveries. With only his or her own interest to sustain an indivdual in her or his task most of the time, the typical Jupiter-watcher of the 19th century, and earlier, preferred finding the 'new' to monitoring the 'old'.

And what to look for? The ultimate observational bias was simply what each observer's interests led her or him to seek. Browning thought almost exclusively about colour. Thomas Webb was interested in variations in albedo. Maria Mitchell concentrated diligently on spots. Edward Knobel was fascinated by large nonaxisymmetric markings. Each of these individuals would probably have considered his or her observations to be objective, but this was impossible. Even a synthesis of their work and that of others, while useful, leaves us without the ability to truly see for ourselves through their eyes.

10.4 Drawing Jupiter

I have deferred until now the subject of drawings made of Jupiter. While in a sense these pictures are a more 'quantitative' source of information, they possess their own inherent set of difficulties when it comes to interpreting them. (Photography remained an astronomical toy during my study period.)

Upon embarking on this subject, I begin with the assumption that each drawing, or at least each body of work by a single artist, is more or less independent. I do so for the previously stated reason that no link can be made between such groupings, with any certitude, other than that the appearance of a feature in a drawing frequently, and not surprisingly, heralds its subsequent appearance soon thereafter in another. The usual absence of commentary on graphic procedure also muddies these waters.

In analysing drawings of the giant planet, it is apparent that the unalterable fact of Jupiter's rapid rotation was obviously a problem to observers. One could start with longitudes near the central meridian and work eastward, but quickly the pace of the planet overtook that of the pencil.

Two approaches were tried. One was a frantic effort to document every detail as quickly as possible. The consequence was often a record of seeing trees but not a forest. (Recall that planetary observers did not notice at first when their theoretical aperture resolution exceeded that allowed them by the sky.) Physically implausible things crept into those drawings that would violate even the crude sense of the wind field of Jupiter existing at the time. In trying to give us a full artistic impression of the planet, meaning was lost.

Others concentrated on rendering the specific features in which they were interested. This led to simpler but unrealistic images. For instance, the drawer often distorted the latitudinal extent of the bands with which she or he was concerned. The result was frequently a drawing with a bloated equatorial region. More space was allotted to that area where the drawer wanted to record more detail. The proper shape of the disc was lost, and positional information became hard to extract because there was no simple functional relationship between latitude and position on the image. The artists who simply omitted the polar regions, or whole hemispheres (to illustrate just their particular point) sometimes give us narrow, but all in all more honest, views of the planet. (Some of these drawings look more like schematics than sketches; indeed, there are allusions in picture captions that suggest that their makers approached them as exercises in technical drawing akin to drafting.)

Technique improved, but what was rendered was still not necessarily useful. Confusion about change—that caused by seeing, that caused by the rotation of the planet, and that which was intrinsically real (and how to deal with it in a drawing), persisted throughout my study period.

Those drawings that actually come to us in published form are the last steps in the editing of the data still available to us today. Initially, the technical barriers involved in printing images (with sufficient resolution to make such an effort worth the page expense) limited presentations to all but the

most-unexpected features. Documentation of discoveries dominates the few
early published pictures of Jupiter even more than it does the written text.
The advent of whole collections of published drawings in the mid-1870s may
not have been based solely on a change of observing philosophy; it also may
have been due to a change in the tolerance of publishers and their medium.
In fact, in an example of an excessive shift of the pendulum, many sets of
a dozen or more Jupiter images made night after night could well have gone
unpublished. Even at the end of my study period, drawings from which no
worthwhile information at all can be extracted were published. The switch
from too few to, arguably, too many pictures in the journals happened over
a short time. (An exception is the chromolithograph, which was used exceedingly sparingly after its development, presumably due to the complicated
nature of the process, and more of which would be welcome.) The desire to
patrol Jupiter by assembling an unceasing series of drawings, and the ability
of the publishing industry to keep up with the result, seem to have come about
at approximately the same time.

Frustratingly, ambiguity in the orientation of prints of Jupiter existed
throughout my study period. By 1645 it was decided that it was acceptable to
draw a planet upside down, and many astronomers began doing so as a matter
of course[5]. Rarely, though, was any caption comment made to the effect that
an image 'hoc est inserne, nam telescopium invertit'[6] after *circa* 1640, when
the modern configuration of the refracting telescope became more popular
than the Galilean arrangement (an arrangement which, at least, produced
a noninverted image)[7]. Fortunately, the northern jet, as a sharply defined
edge of the north tropical zone, is so consistently striking and latitudinally
invariant that it can be used in almost every case to orient artwork properly.
(Most images of Jupiter were printed literally as they would appear through
the optical system in use for observation.) The fact that it is possible to
orient unambiguously these images and, furthermore, identify certain bands
and their symmetry speaks well for our current understanding of the dynamics
of the visible jovian atmosphere.

The selection of drawings published, and the loss of detail unavoidable
in the publishing process, limits their interpretive usefulness today. They
must be considered ancillary to the redacted record, though still an important resource—especially for identifying features that were not recognized *as*
features at the time a drawing was made. The drawings are most useful in
providing 'background' information concerning the general appearance of the
belts and zones. Albedos and breadths of the belts and zones (the fundamental underlying morphology of the visible planet) were not often of primary
interest during the history of Jupiter under discussion. (The equatorial zone
reddening of 1870 was the first major exception to this.) Notwithstanding,
the belts and zones do appear in the drawings. These 'inadvertent' data,
sometimes more objective than those presented with the purpose of marking
a discovery or the ability of a telescope, may be the major legacy of this
artwork.

10.5   The Histories of Selected Jovian Features

Concerning those features important in modern images of Jupiter, of which I have traced historical sightings, I begin my discussion with the southern hemisphere of Jupiter and work northward.

The first records of STZ spots are those written by William Dawes, during the Spring of 1849, and the drawings produced by William Lassell, beginning in March 1850. While scattered reports of southerly white spots exist, some even in illustrations from before this time, few seem meant to be placed farther south than the south equatorial belt. Previously, I mentioned the difficulty of observing temperate features and the fact that these latitudes were considered unreliable for rotation-timing. Thus, neither the ability nor desire to look for phenomena in the area existed through much of this time. (Lassell's spots were the first major features on Jupiter to be discovered using a reflecting telescope.)

The key to the discovery of Lassell's spots is their appearance in irregular strings. Lassell's and Dawes' recognition of this unusual aspect cleared the way for many observers to see them in ensuing decades. During the first part of the 20th century, that period summarized by Bertrand Peek[8], they seem nearly to have been forgotten. Obviously, their transient nature caused them to be intrinsically less interesting than longer-lived features, such as the Great Red Spot or White Ovals, where the history of the phenomenon may be that of the longevity of a single feature.

Peek[9] does not mention any possible precursors to the White Ovals of 1938. I have discovered candidates drawn by three observers, in 1872, 1873 and 1874, respectively. In each of these sets of observations, the depictions seem quite different; but then the modern Ovals evolved very quickly, and so it is not possible to say whether these were the record of the evolution of a single feature or several unrelated ones. The disappearance from the study-period data of major white STB spots altogether, after this initial tentative interest, is disturbing, though. I must assign a fairly low confidence level to the identification of any of these observations exactly with the morphology described, over 50 years later, as White Ovals.

As south tropical zone disturbances are fairly long lasting, it is not surprising that they were the first of the regularly occurring 'disturbance' phenomena to be acknowledged. In comparing my research to that of Peek[10], I concur that the first records are the representations made by Dawes in 1857, though while Peek cites the 28 November rendition, I can detect the disturbance 'head' in the version rendered one night earlier. Peek includes a hitherto unpublished drawing, inked by Samuel Schwabe on 13 November 1859, that shows a probable disturbance in the STrZ. Again, I agree with Peek's interpretation of Knobel's 7 and 24 March 1874 pictures, as depicting a STrZ disturbance, but would move its first appearance up to the 6 March version. I can follow the disturbance's development in Knobel's artwork until the beginning of May.

The first highly credible record of the Great Red Spot Hollow that I have found is Bond's February 1848 observation. The first drawing (since Schwabe) is, with less certainty, Piazzi Smyth's, included in his 1856 Tenerife series, a rendition made in September 1856. Shortly thereafter, the Hollow is recorded several times in a sequence of drawings made by Dawes to illustrate Lassell's spots, in December 1857. Angelo Secchi also drew it that month. In another set of drawings produced to illustrate yet another unrelated phenomenon, Joseph Baxendell and colleagues demonstrated the presence of the Hollow in early 1860.

Two of Browning's drawings (of 1869 and 1870) show probable Hollows. The latter is confirmed by a similar concurrent representation by Webb. These observations straddle those of Alfred Mayer's ellipse.

Note that all these observations (except the ellipse) are of the Hollow and not the Spot. Also, none of its observers wrote of it or, apparently, attached any special attention to it. Indeed, here was a situation where we benefit from the truly more objective nature of certain drawings.

The elliptical feature of Mayer (and Joseph Gledhill), seen in late 1869 and early 1870, is the first picture I have seen that reasonably portrays a Spot-like feature (but no Hollow). Mayer's and Camille Flammarion's descriptions of its colour make it likely that this was really the Great Red Spot. It was also unique in that it was *identified* as a noteworthy feature, but nevertheless only considered transient, it seems, because of the absence of attempts at the time to discover precursor observations (and its rapid disappearance from the literature once it had faded).

From December 1872, The Great Red Spot Hollow probably appears in a depiction created under the direction of Laurence Parsons. If this and other portrayals by Browning, Webb etc were recognized, there is absolutely no evidence that they were. Drawing does not constitute recognition. No attempt was made to connect its appearance with Mayer's ellipse, still thought of as a single apparition's anomaly until George Hirst's 'fish' surfaced in 1876.

It was not until 1878 that a spot, this time definitely red, appeared to observers at the modern GRS latitude on Jupiter. It dominated the planet's surface between 1879 and 1882 as its largest and darkest feature. While the oval, then formally named the Great Red Spot, indented the SEB(S) (halving its width at its maximum latitudinal extent), the connection between its behaviour and that of the previously drawn Hollow still was not made immediately.

Other studies accessing unpublished work have identified other potential pre-1878 manifestations of the GRS (principally, that by William Denning in 1898[11]). Their method involved locating records of any features that had the appropriate rotation time assumed for the Spot. Some difficulties inherent in this method already have been addressed. Still, it is not my intention to extend this work significantly.

The Red Spot and Hollow finally were equated. The Hollow became thought of as but a peculiar form of the spot feature. Today, we recognize that it is the Hollow that is the physical morphological form at those latitudes. It is an anticyclonic oval that may from time to time expose a red chromophore brought to its visible surface through some not yet understood mechanism. It is the Hollow that is the long-lived feature, which, perhaps can extract enough energy from surrounding wind jets to sustain itself.

Yet, historically, the aesthetically appealing Spot has been considered the intrinsic feature. The post-1878 historical search for the Great Red Spot literally has been coloured by this fact. If the Hollow had been recognized as a real feature before 1878 and not, in hindsight, as a phenomenon associated with the Great Red Spot, we might have a much more detailed record of it. This would benefit us in learning its longevity. This is an important question that cannot be answered: while the Hollow appeared in some pictures, primarily after more rigorous attempts at objective artistry were made during the last quarter of the 19th century, in how many more drawings would we see it if it, too, had been recognized as a sort of 'spot'? The case of the written record is more extreme. Again, what was perhaps present, but not recognized as significant, was not recorded. A historical search for *red* features before about 1870 does not yield the whole story of this vortex. In that sense, the GRS is a 'red herring'.

Continuing my review of major jovian morphology northward, I have found several examples of south equatorial belt disturbances before that of 1919–20 reported by Peek[12]. I have found evidence for an SEB disturbance, of at least two years' duration, in the work of John Birmingham, beginning late in 1869. Étienne Trouvelot and Parsons drew an SEB disturbance in 1873. Fedor Bredikhin and Knobel did so in early 1874. (It looks like a new disturbance forming, but this is far from clear.) The 1874 disturbance also is incorporated in the artwork of Flammarion and Oswald Lohse.

There are no reports from observers in the southern hemisphere, who were in a position to monitor both the STrZ and SEB disturbances that had been taking place in the mid-1870s. This was a new sort of phenomenon their colleagues in the northern hemisphere were just starting to record. There is no reason to suppose that this kind of activity would be recognized coincidentally as 'features' by them at this same time. Unfortunately, even during these years, most of the published representations come to us from northern observers.

Because of the nature of the northern hemisphere of Jupiter, my discussion of it will be brief. Equatorial plumes, like several other phenomena, appeared in the observations of Dawes beginning in 1851. Bond possibly preceded him by a few years. Others observed them during the 1850s; the first drawing of plumes, though, was Dawes' portrayals of late 1857. From then on they turned up frequently and were quickly considered commonplace.

The discovery of diagonal features extending from a convective source in the NEB has been described in detail. The story of the notable example known as the 1860 Oblique Streak was presented as a study in how a phenomenon could become the subject of great fascination and then, once its uniqueness was lost through the discovery of additional—albeit lesser— examples, soon be relegated to the type of marking often included in sketches but rarely generating comment.

Documentation of northern 'barges' appears to go all the way back to Wilhelm Beer and Johann Mädler during the 1830s. Earlier, it was discussed why such features should have been among the earliest identified. Furthermore, even 18th-century reports of dark northern hemispheric spots may well have been those of 'barges'. Resolution did not permit observation of their distinctive elongated shape and their truly great contrast with their surroundings. In other words, they were not linked to a single recurring morphology at a particular latitude.

'Barges' appeared in descriptions of the planet intermittently before being formally named—after my study period. Because of their small extent, though, they were difficult to render in paint or crayon; their appearance was more easily put into words. By the 20th century, 'barges' were expected on Jupiter. But because of the astronomical public-relations fiasco brought on by Percival Lowell (1855–1916) and his erroneous interpretation of martian 'canali' (their construction by an advanced civilization), the actual term went out of favour. The introduction of 'canal language' into jovian astronomy was to be avoided at all cost[13]! (I have retained the term, but in deference to such concern, I have used quotation markings surrounding it throughout this text.)

Peek cites 1880 as the year of the discovery of what we now call the northern jet[14]. Here, however, he must mean Denning's actual determination of a wind velocity by timing features flowing within it, for in fact, the sharply delineated albedo contrast caused by the straight jet can be seen beginning with some of the very first Jupiter illustrations, often when nothing else familiar is present. It was used throughout my analysis of Jupiter drawings as a starting point for establishing orientation and then latitude.

Significant, too, are the features that I do not see. I can find no reliable evidence in the historical data for the small brown features in the north tropical zone[15]. Here, if the limitation was optical, it must have been in size and contrast, for still more poleward features were identified in the south.

Likewise, I can find no counterpart for the ephemeral 'little' red spots found at a north latitude symmetrical to the Great Red Spot's in the south. These are the subject of recent interest because of their distinctive chromophore, which is seemingly the same as that of the GRS[16].

10.6   Were there Impacts on Jupiter before Comet Shoemaker-Levy 9?

Before the last Shoemaker-Levy 9 fragment had fallen, it was already proposed that a search be made of historical records of jovian spots. The goal is to find descriptions or depictions of similar spot morphologies, if they exist, that were ignored or attributed to meteorological causes in their time. The discovery of spots similar to those produced by SL-9 would suggest that they, too, might be produced by impacts. In turn, this information would tell us something about the distribution of small bodies in the solar system.

The features created by Shoemaker-Levy 9 on Jupiter were almost immediately and spontaneously described as 'spots' by observers all over the world (e.g., Internet Exploder transmissions of July, 1994). Early journals and observing records are *filled* with reports of 'spots' on Jupiter. We read of 'light spots', 'dark spots', 'white spots', 'black spots', 'bright spots', 'dusky spots', 'small spots', 'large spots', 'minute spots', 'enormous spots', 'circumscribed spots', 'round spots', 'isolated spots', 'distinct spots', 'intense spots', 'fine spots', 'hard spots', 'well-defined spots', 'conspicuous spots', 'peculiar spots,' and 'remarkable spots'. These subjective adjectives give little hint about the morphology of the spot being described.

Why were jovian spots described so superficially? In many cases, telescopic resolution and atmospheric seeing prohibited any more detailed discussion. But in many others, the reason lay in the fundamental purpose and methodology of observing the spot:

(i) Descriptions of jovian spots were not *valued*. Well into the eighteenth century, astronomy was considered a science of measurement. What could not be measured, was not thought significant. Morphological descriptions were by definition qualitative. The only physical measurements that could be made concerning Jupiter were those of its oblateness and its rotation period. Throughout most of the last four centuries, spots on Jupiter were observed to determine the rotation period of the planet and for no other purpose. Usually *light* spots were favoured for rotation timing. Even when dark spots were used, those with a minimum of structure were preferred. An ideal spot for rotation timing was one with high contrast and distinct, simply defined edges, i.e. a black dot.

(ii) Those who did examine dark spots often misinterpreted them. Dark spots, which concern us most in the search for Shoemaker-Levy 9 analogue spots, represented the *absence* of a feature to many observers of Jupiter. They were thought to be gaps in a pervasive jovian cloud deck, gaps that allowed the observer to glimpse the fixed 'surface' of Jupiter below. At best, they were important because the measurement of their transit times might hopefully yield the period of the supposed solid bulk of Jupiter. When peeking through a locked door, why spend time studying the *keyhole*? Detailing the morphology of a 'hole' would have been considered of little use to generations of Jupiter-watchers.

(iii) The latitude of a spot is often missing from a reported observation. There was no apparent need to specify latitude! Jupiter's differential rotation was not widely considered. Therefore, latitude was irrelevant: if the purpose of observation were to time the rotation period of Jupiter, the result yielded by any spot should be the same. If it were not, there were plenty of observational and data-reduction errors that could be blamed. Location on the planet is important because spots commonly appear at certain latitudes. No one has ever suggested that these spots have other than an internal origin. For instance, a 'chain' of spots on Jupiter was observed as early as the 1840s. But these were intrinsic Lassell spots, not 'comet spots'.

(iv) Rotation timing often took the place of other observations of Jupiter. For instance, most observations were made in the equatorial regions of Jupiter, to avoid foreshortening at the north or south limb. Spots that appeared in the polar regions might be overlooked altogether. Even into more modern times, when keen morphological interest in Jupiter arose, polar features were missed, or not looked for, because of the lack of contrast there. There are few descriptions of any features poleward of the first temperate bands on Jupiter.

(v) Features that have been historically documented are preferentially in the southern latitudes. As the historical observers found little of interest in the 'quiet' northern hemisphere of jupiter, attention focused on the equatorial zone, south equatorial belt, south tropical zone, and south temperate belt and zone, where most long-lived spots and zonal disturbances seem to appear.

(vi) It is difficult to continually monitor Jupiter. Just as observational coverage over the area of Jupiter is not complete, it is also incomplete over time. Most of the observers who chose to monitor Jupiter did so from northern latitudes. For those years during which Jupiter was at negative declinations, high air masses prohibited truely good observing conditions—even at opposition. Compared to the behaviour of an object like the Moon, Jupiter's rapid rotation has always been an obvious problem for visual observers. Studying the entire disc, or worse yet, trying to draw it, Jupiter-watchers found themselves in the same position as an artist at an easel attempting to paint an ornate but indistinct passing locomotive. In the astronomical case, the situation was even worse. Instead of a continually clear, moving image, astronomers were faced with randomly timed 'snapshots' of the planetary disc. This was due to variable and erratic blurring of the image caused by bad seeing. There simply was not time to doccument every detail at every longitude.

There remains one technique that may yet yield Shoemaker-Levy-like spots in the historical record. In 1994, those who observed the SL-9 spots through small ground-based telescopes noted a unique character associated with them, apart from their morphology. Normally, jovian cloud features show the strongest contrast near the central meridian. As they rotate toward the limb, the increased mass of overlying hazes in our line of sight reduces contrast, and the spots tend to disappear before they cross the limb or terminator. This was not true for the SL-9 spots. The material in the SL-9 spots was deposited in the jovian stratosphere. At the limb, an increased mass of spot material in the line of sight caused limb contrast *enhancement*. The only other jovian feature that does not show relative limb contrast loss is a satellite shadow, which displays constant contrast as it transits the planet.

The characteristic described above would have been detectable by historical observers of Jupiter. If a further search of the historical record yields reports of a dark jovian spot with unusual limb properties, that spot may be considered a pre-SL-9 impact candidate.

10.7 Barriers to Interpretation

The characterization of Jupiter presented here is mostly a familiar one. The picture of Jupiter left to us in ink and paint from past centuries, taken as a whole, includes details routinely visible today. Yet it is impossible to conclude from this what one of the astronomers from the past, who has figured in this story, might record or draw if transported to our day and confronted with Jupiter. It is probable that the image so constructed would be similar to that which such an observer might have produced in his or her own day, if that individual were allowed to use similar equipment and media. It also seems likely that, subjective as these views were, such a modern report would *differ* greatly if the observer also was made privy to what knowledge now exists about the physical external and internal mechanisms at work on the planet and its smaller-scale morphology.

The historical vision of Jupiter seems to evidence something that is not unexpected: that while the way in which we view the planet evolves, on the scale of human history, what occurs there does not. When we strip away the barriers to interpretation that are a function of the data's place in time, we see that what has happened on Jupiter during the interval of our scrutiny has done so repeatedly and consistently. By extrapolation, what has happened on Jupiter will happen again, as Jupiter responds to a complex, but real, large-scale stability. Prediction of events on Jupiter is possible, though the giant planet at first glance seems so unordered on its facade. Jovian meteorology can and will continue to progress because, over the lengths of time we are ever likely to be practically concerned with, on Jupiter it can be said that the past is prologue.

But what of the conundrum of the tautology, implicit in this assessment and mentioned in my introduction? The interpretation above is made perforce because of the inability to see Jupiter through 16th, 17th, or 18th-century eyes. Therefore, in summary, let me restate those things that must qualify my analysis. The physical scientist would call these biases on the available data. The historian might call them mechanisms through which historical facts— not in actuality fixed forever, but malleable, evolutionary, and dynamic—were transformed into their present forms. Both statements are characterizations of those things, hopefully made manifest in this study, that obscure and fog our view into the past and of which we must be cognizant in making use of this data.

(i) *Resolution.* Before any celestial phenomenon can be observed, it must first be resolvable by the combination of telescope and human eye. These are technical and physiological limitations. We cannot hope for much improvement in the latter. However, means to effect the improvement of the former continually were extended during my study period, as was understanding of the theoretical meaning and importance of resolution.

(ii) *Seeing conditions.* Further restricting what can be seen and/or resolved are seeing conditions, which depend on geography, meteorology, and time. Planetary astronomers, for whom choice seeing was most critical, were nonetheless slow to optimize these variables.

(iii) *Cognitive factors.*† Hearing and sight are quite different from sound and light. The role of the human brain in recording data about Jupiter poses a dilemma: it is an *individual* brain that describes the telescopic view at a given eyepiece at a given instant, not a pair or team of brains. The physiological and psychological differences that provide the variety of the human species also assure that we cannot be certain that any two observers actually 'see' the same thing. This is no more true than in the analysis of colour, colour bias and colour blindness being well documented. Colour determination remains an art.

(iv) *Latitude.* A person's location on the globe (and desire and ability to change that location) determines whether he or she will periodically see Jupiter high in the sky through low air masses, with all the observational benefits this entails, alternating between times when the planet hovers near the horizon, with the corresponding detriment to observing. It also will dictate what the phase of this oscillation will be and its amplitude.

(v) *Instrumentation.* That there are differences between observing Jupiter with a reflecting telescope and a refracting telescope (and between two instruments of even the same optical design, aperture and focal length) was accepted in the era of the visual observer. Nevertheless, a concordance was never reached on what those differences were, nor was the subject even satisfactorily discussed in the literature. Eyepieces were mentioned still more rarely.

(vi) *Observing Technique.* Did the astronomer work at dawn or dusk? Did she or he work for long periods? Was he or she prone to pressing the limits of seeing by using high magnification? Did she or he put up with uncomfortable viewing conditions? Was the observer patient?

(vii) *Experience.* Only when observers came along who studied Jupiter telescopically, for long and relatively uninterrupted periods, could eyes be 'trained' to interpret as having meaning many of the forms found on Jupiter, effectively increasing practical resolution. That such persons were not immediately available after the invention of the astronomical telescope was not so much because of any technological limitation but rather for sociological reasons.

(viii) *Objective.* Human resources and time are finite. Whether an observer's object is to document everything that appears on the jovian disc at a certain hour, or is to concentrate on the behaviour of targeted and well

---

† Psychiatrist and astronomer William Sheehan writes much more about the role of perception in planetary observation in his seminal 1988 *Planets and Perception* (Tucson, AZ: University of Arizona Press).

followed features (to the exclusion of global context), profoundly affects the resulting end record and the amount of quality information available for particular jovian latitudes and longitudes.

(ix) *Purpose*. Similarly, if an observer, recognizing the restriction posed above on effectively witnessing everything that shows up on Jupiter, has directed her or his attention to one or more specific goals, the full account produced by that observer also will reflect other aspects of the planet, but to a greater or lesser extent, depending on the goal. An individual concerned with timing the rotation of Jupiter will portray the planet in its longitudinal dimension much more carefully than in its latitudinal dimension. A person looking for the solid 'surface' of Jupiter will sacrifice positional information for contrast. Those would-be discoverers who search for the new will neglect the old. Others bent on writing the history of the development of existing features may themselves overlook newly appeared ones.

(x) *Language*. By language I mean a distinction between English, German, French and others and the fact that certain of these languages may better suit the discussion of Jupiter. I also mean, though, nuances caused by using jargon, technical phrases, artists' expressions, the almost slanglike terms (because of the specific and sometimes unconventional definitions used) adopted by the planetary science community or, merely, flowery prose such as that which was popular during the 19th century. (The reader has probably noticed that few of the quotations from this time that I have used in this work sound like modern scientific writings.) Such choices in speech affect which aspects of a communication 'come through' to us and which are 'muted'. The use of individual words, even, is significant because they easily can be influenced by particular preconceptions or working models within whose boundaries the author's mind works. Yet these may go unstated explicitly.

(xi) *Artistic ability*. Drawing Jupiter, as we have seen, adds a step to the documentation process that is absent when the observer relies on the written word for communication. No doubt some observers were incapable of satisfactorily drawing the planet and had to stand on whatever powers of written exposition they possessed. We have little idea who fitted into this category and who did not draw Jupiter, or happened to use description over rendition, merely by choice.

(xii) *Artistic style*. Intentionally or unintentionally, the medium and manner by which a planetary artist portrays Jupiter lend themselves to the clarification of some features, and the virtual obliteration of others, when he or she reduces the planet to two dimensions.

(xiii) *Drawing technique*. This 'filter' not only includes technical limitation of the drawing process but also whether finished works were produced in stages, from notes or rough drafts. It further includes whether one or more persons participated in each stage, thereby adding other minds to the chain of interpretation.

(xiv) *Reproduction*. Our view of jovian artwork (and, for that matter, written text about Jupiter) from the past is controlled by how well it is preserved. For published material, such as the data base for this study, both the means of reproduction *and* the preservation of the reproduced version serve this function. Bounded by the laws of information theory, each reproduction step means lower information content (with a variable degree of degradation between steps). This historical entropy continues today, depending upon the quality of the microfilm, microfiche, microcards, cameras and photocopying machines by way of which I have compiled my working data set. Likewise, by examining *this* work (in printed form or in microfilm facsimile), the reader does not see exactly the same Jupiter, or typeset words about it, that I saw.

(xv) *Audience.* Was a rendering made as a true representation of Jupiter or was it supposed to be, foremost, appealing to the eye? Similarly, in writing, was the intended audience other members of the planetary-science community, scientists at large, or the public (who had an increasing appetite for popularized science during the years that were my study period)?

(xvi) *Impartiality.* Did the observer have any particular theory or hypothesis that she or he was attempting to defend or promulgate? (Here I speak of *intentional* inobjectivity.)

(xvii) *Objectivity.* While almost always the stated goal of attempted objectivity led to self-consistency, it never could lead to a truly judgment-free translation of the appearance of Jupiter to the printed page. Epistemologically, a fundamental limitation of all history applies equally well here: the published knowledge that will be transmitted to the future still will have written 'between the lines' the thoughts and perspectives of the authors, when it is read years, decades, or centuries later at a time when our opinions of what constitutes this planet Jupiter have radically changed. Then the ability to reach these persons' thoughts from the mere messages they bequeath will be necessarily impaired.

Endnotes

[All titles are written in full, with the exception of '*Mon. Not.*' for the *Monthly Notices of the Royal Astronomical Society*.]

1. Firmstone E 1872 The planet Jupiter *Winchester and Hampshire Scientific and Literary Report of Proceedings* 81
2. Knobel E 1873 Note on Jupiter, 1873 *Mon. Not.* **33** 474
3. King H 1955 *The History of the Telescope* (Toronto: General Publishing)
4. Fritsch 1800 Wahrnehmungen der *Sonnenflecken,* der Venus, und des Jupiters, mit einem $2\frac{1}{2}$ f. Ramsdenschen Fernrohr, imgleichen eine astron, Nachricht *Astronomisches Jahrbuch* 110
5. Van Helden A 1974 The telescope in the seventeenth century *Isis.* **65** 38
6. Bianchini F 1737 *Veronensis Astronmicae ac Geographicae Observationes Selectae* (Verona: Manfredi)
7. Van Helden A *Op. Cit.*
8. Peek B 1958 *The Planet Jupiter* (London: Faber and Faber)

9. *Ibid.*
10. *Ibid.*
11. Denning W 1898 The red spot on Jupiter, and its suspected identity with previous markings *Nature* **58** 331; Denning W 1898 The Great Red Spot on Jupiter *Mon. Not.* **58** 488
12. Peek B *Op. Cit.*
13. *Ibid.*
14. *Ibid.*
15. Beebe R and Terrile R 1979 Summary of historical data: interpretation of the pioneer and voyager cloud configurations in a time-dependent framework *Science* **204** 948
16. Beebe R and Hockey T 1986 A comparison of red spots in the atmosphere of Jupiter *Icarus* **67** 96

Appendix A

Oppositions of Jupiter

Observing apparitions are centred on the dates of Jupiter's opposition. Near opposition, the superior planet is closest to us, and presents its maximum angular diameter when viewed through the telescope eyepiece. The 'barges' of Johann Mädler, Lassell's spots, the Oblique Streak, the 1870 equatorial zone reddening, Alfred Mayer's ellipse, George Hirst's 'fish', and Carr Pritchett's Great Red Spot all were discovered within days of opposition.

This table was constructed by Siobahn Morgan, using *Astronomy Lab* (version 1.13) by Eric Bergman-Terrell. The interval between oppositions is the synodic period of revolution of Jupiter (399 days), which is a function of the sidereal periods of revolution for both Jupiter and the Earth†. Dates are Gregorian; the Julian calendar was used in Britain and its colonies until September 1752.

The jovian plane of revolution about the Sun is higher in our night sky during the winter than during the summer. Oppositions occurring in the northern hemispheric winter are italicized. These represent 'windows' of optimum observing conditions for Jupiter. It is interesting to note that both Galileo Galilei and William Herschel coincidentally observed Jupiter during such 'windows'.

† $P_{synodic} = (P_{Jupiter} \times P_{Earth})/(P_{Jupiter} - P_{Earth})$

Bishop R 1996 *Observer's Handbook 1997* (Toronto: Royal Astronomical Society of Canada)

Seventeenth Century

7 March 1601
7 April 1602
8 May 1603
9 June 1604
14 July 1605
20 August 1606
27 September 1607
2 November 1608
8 December 1609
10 January 1611
10 February 1612
12 March 1613
11 April 1614
13 May 1615
14 June 1616
20 July 1617
26 August 1618
2 October 1619
7 November 1620
12 December 1621
14 January 1623
14 February 1624
16 March 1625
16 April 1626
17 May 1627
19 June 1628
25 July 1629
31 August 1630
8 October 1631
12 November 1632
17 December 1633
19 January 1635
19 February 1636
20 March 1637
20 April 1638
22 May 1639
24 June 1640
30 July 1641
5 September 1642
13 October 1643
17 November 1644
22 December 1645
23 January 1647
23 February 1648
24 March 1649
24 April 1650

27 May 1651
29 June 1652
4 August 1653
11 September 1654
18 October 1655
22 November 1656
26 December 1657
27 January 1659
27 February 1660
29 March 1661
29 April 1662
31 May 1663
4 July 1664
9 August 1665
16 September 1666
23 October 1667
27 November 1668
31 December 1669
1 February 1671
3 March 1672
2 April 1673
3 May 1674
5 June 1675
9 July 1676
14 August 1677
21 September 1678
28 October 1679
2 December 1680
5 January 1682
5 February 1683
7 March 1684
6 April 1685
8 May 1686
10 June 1687
14 July 1688
20 August 1689
26 September 1690
3 November 1691
7 December 1692
9 January 1694
10 February 1695
11 March 1696
11 April 1697
12 May 1698
14 June 1699
20 July 1700

Eighteenth Century

26 August 1701
3 October 1702
9 November 1703
13 December 1704
15 January 1706
15 February 1707
16 March 1708
16 April 1709
18 May 1710
20 June 1711
25 July 1712
31 August 1713
8 October 1714
14 November 1715
17 December 1716
19 January 1718
19 February 1719
21 March 1720
20 April 1721
22 May 1722
25 June 1723
30 July 1724
5 September 1725
13 October 1726
19 November 1727
22 December 1728
23 January 1730
23 February 1731
25 March 1732
25 April 1733
27 May 1734
30 June 1735
4 August 1736
11 September 1737
18 October 1738
24 November 1739
27 December 1740
28 January 1742
28 February 1743
29 March 1744
29 April 1745
1 June 1746
5 July 1747
9 August 1748
16 September 1749
23 October 1750

28 November 1751
31 December 1752
1 February 1754
4 March 1755
2 April 1756
4 May 1757
5 June 1758
10 July 1759
14 August 1760
21 September 1761
29 October 1762
3 December 1763
5 January 1765
6 February 1766
8 March 1767
7 April 1768
8 May 1769
10 June 1770
15 July 1771
20 August 1772
26 September 1773
3 November 1774
8 December 1775
9 January 1777
10 February 1778
12 March 1779
11 April 1780
13 May 1781
15 June 1782
20 July 1783
25 August 1784
2 October 1785
8 November 1786
13 December 1787
14 January 1789
14 February 1790
17 March 1791
15 April 1792
17 May 1793
19 June 1794
25 July 1795
30 August 1796
7 October 1797
13 November 1798
18 December 1799

Nineteenth Century

1 January 1801	8 April 1851
19 February 1802	8 May 1852
22 March 1803	10 June 1853
21 April 1804	15 July 1854
23 May 1805	21 August 1855
25 June 1806	26 September 1856
31 July 1807	3 November 1857
5 September 1808	8 December 1858
13 October 1809	*11 January 1860*
19 November 1810	*10 February 1861*
23 December 1811	*13 March 1862*
24 January 1813	12 April 1863
24 February 1814	13 May 1864
26 March 1815	15 June 1865
25 April 1816	20 July 1866
27 May 1817	26 August 1867
30 June 1818	2 October 1868
5 August 1819	8 November 1869
11 September 1820	13 December 1870
18 October 1821	*15 January 1872*
24 November 1822	*14 February 1873*
28 December 1823	*17 March 1874*
28 January 1825	17 April 1875
28 February 1826	17 May 1876
30 March 1827	20 June 1877
29 April 1828	25 July 1878
1 June 1829	31 August 1879
5 July 1830	7 October 1880
10 August 1831	13 November 1881
16 September 1832	18 December 1882
24 October 1833	*20 January 1884*
29 November 1834	*19 February 1885*
2 January 1836	*21 March 1886*
2 February 1837	21 April 1887
4 March 1838	22 May 1888
4 April 1839	24 June 1889
4 May 1840	30 July 1890
5 June 1841	5 September 1891
10 July 1842	12 October 1892
15 August 1843	18 November 1893
21 September 1844	*22 December 1894*
29 October 1845	*24 January 1896*
3 December 1846	*23 February 1897*
6 January 1848	25 March 1898
6 February 1849	25 April 1899
9 March 1850	27 May 1900

Index

Thomas Hockey studied astronomy, history, and philosophy at Peoria High School, the Massachusetts Institute of Technology, and New Mexico State University. He has been an Instructor at the College of Wooster, Faculty Associate at Arizona State University, and Visiting Professor at Shaanxi Teachers' University (PRC). He was a Fellow in the NASA JoVe Program. Today, Hockey is Associate Professor of Astronomy at the University of Northern Iowa and Secretary of the American Astronomical Society's Historical Astronomy Division. He has presented talks and papers on six continents. Hockey is the author of three books and is the producer of the movie 'Clyde Tombaugh and the Discovery of Pluto'. Thomas and Anne Hockey, plus three children, live in Cedar Falls, Iowa, in a nineteenth-century Victorian house decorated with Jupiters.